THE
HUMAN BODY

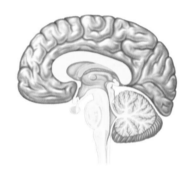

THE
HUMAN BODY

AN ESSENTIAL GUIDE TO HOW THE BODY WORKS

GENERAL EDITOR:
JANE DE BURGH

First paperback edition published in 2007 for Grange Books
Reprinted in 2009, 2010

An imprint of Grange Books Ltd
35 Riverside
Sir Thomas Longley Road
Medway City Estate
Rochester, Kent
ME2 4DP
www.grangebooks.co.uk

ISBN 978-1-84013-880-1

Editorial and design by
Amber Books Ltd
Bradley's Close
74–77 White Lion Street
London N1 9PF
www.amberbooks.co.uk

Project Editor: Marie-Claire Muir
Design: Neil Rigby at Stylus Design

Printed in Thailand

PICTURE CREDITS

Photographs courtesy of Ralph T. Hutchings
All artworks © Bright Star Publishing plc

This material has been edited and condensed from material previously published in the partwork, *Inside the Human Body*.

CONTENTS

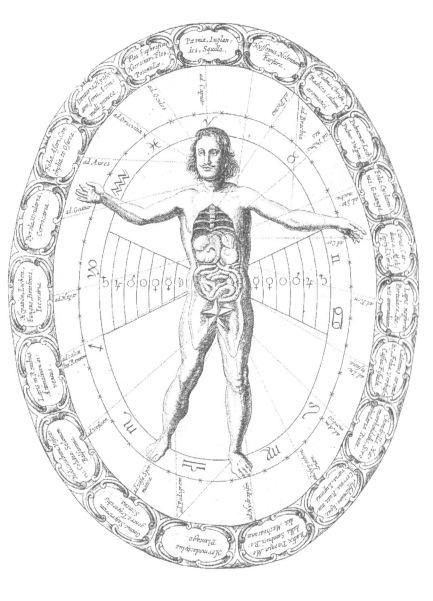

Introduction

Anatomy is the oldest-known medical science. Our fascination with our bodies and how they work, why they go wrong, and what to do to heal them is boundless. Throughout history, countless theories, mostly erroneous, explaining anatomy and physiology have been dreamt up by all kinds of physicians, surgeons, quacks, witchdoctors, alchemists, faith-healers, astrologers, and charlatans, who, in their day, were often well respected and highly paid professionals.

Despite this catalog of bad practice, the history of medicine is punctuated by brilliant discoveries and truly visionary thinking that has, against all the odds, hauled us into the modern era of medical science. Hippocrates, "the father of medicine," was a physician on the Greek island of Cos in the fifth century B.C.E., and he is undoubtedly the most famous and recognizable figure of them all. He established a specialist body of physicians who were governed by a strict code of ethics, and who employed observable scientific methods in their research. This laid the foundation for modern medical practice.

THE FOUR "HUMORS"

Hippocrates' work had a profound influence on medicine, and his ideas were enthusiastically expanded upon by doctors in the centuries that followed. However, Hippocrates' knowledge of internal medicine was limited due to the lack of dissections, and his theories on anatomy and disease were factually inaccurate. He believed that four "humors," (black bile, yellow bile, phlegm, and blood) governed human health and that any illness was a result of imbalances between them. He also insisted that the the relationship between patient, physician, and disease was important in the diagnosis and treatment of illness, a novel philosophy at the time, given that diseases were still presumed to be punishments imposed by the gods. In this sense Hippocrates could also be said to be the father of holistic medicine.

With the exception of the monks, who grew herbs and plants with some genuine medicinal properties, factual inaccuracy was the trademark of medicine and anatomy during the Middle Ages. The "humors" theory was still widely held to be true, and Christian and Islamic religious belief was highly

Left: In this seventeenth-century diagram, the human body represents the world in microcosm, which is described as a living organism with metabolic processes.

influential on medical theory. All sorts of theories, such as bloodletting, draining "noxious fluids" from the body, or encouraging "excess fluids" to move freely around the body, were commonly put into practice, often accompanied by apothecaries' potions, which contained such bizarre and infamous ingredients as newts' tongues and worms' livers.

With the arrival of the Renaissance in Italy in the late fourteenth century, medical science moved forward. The rediscovery of Classical learning encouraged physicians to reapply scientific methods to medical research and leave behind the influence of religion and superstition. Great names from the period, such as Leonardo da Vinci, put forward new ideas. He believed that in order to treat disease it was necessary to first learn about the body and its processes, learning that could ultimately only come about through the dissection of human cadavers. Dissection was not, however, a new idea. Claudius Galen, a highly influential second-century physician, had dissected animals and had assumed that human anatomy followed the same patterns, an idea that became accepted wisdom for more than 1,500 years. But by the sixteenth century, the anatomist Andreas Vesalius demonstrated that Galen was wrong and revealed previously unknown anatomical structures in his book *de Humani Corporis Fabrica* (Fabric of the Human Body), in 1543. Other pioneering work recording what had been discovered was conducted by da Vinci and Vesalius, who sought to accurately represent anatomical structure through detailed diagrams and illustrations.

BLOOD CIRCULATION

Still, these ideas and methods were controversial and often dismissed. In 1628, the English doctor, William Harvey, stunned the medical world when he published *An Anatomical Disquisition on the Movement of the Heart and Blood*. In this book, Harvey showed that blood circulated around the body, and he further proposed that the heart pumped blood through arteries. He also realized the significance of the valves of the heart in controlling the flow of blood. Although Harvey's ideas were considered outlandish, this scientific method of research was again proved to be the way forward. His findings were confirmed by the invention of the microscope in the late seventeenth century. For the first time in history, scientists could observe more than the naked eye would allow.

MODERN MEDICINE

By the end of the nineteenth century, many of the practices and procedures we now take for granted were coming to the fore. Crude anesthetics were

In this engraving by English artist and satirist William Hogarth, a body is dissected in an operating theater. The depiction of the doctors and the treatment of the cadaver reflect the public's negative attitude toward anatomists and the practice of dissection.

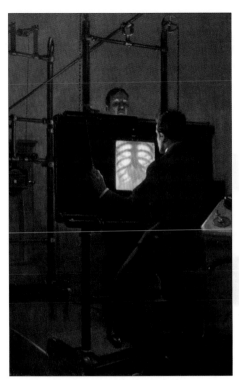

developed by James Young Simpson, antiseptics were pioneered by Joseph Lister, and in 1896, Wilhelm Röntgen amazed the world with a new invention—the X-ray machine—that allowed the internal examination of the body without the need for surgery. Other groundbreaking work by scientific figures, such as Louis Pasteur, who established the link between germs and disease, and Karl Landsteiner, who discovered the four main blood groups, paved the way for much more complex surgery, such as organ transplants. The 1980s saw the advent of computed tomography (CT) scans, which use thin X-ray beams that rotate around the patient to gather information. The data is collated by a computer to create an image of the internal structures of the patient's body, showing the relative location of the structures in cross-sectional orientation. Today, surgeons can perform what would have been a miracle only 200 years ago.

Dr. Diocles' radio stereoscope (circa 1926) presents a 3D image of the subject.

PRACTICING ANATOMY

Dissection of human cadavers, a key element of contemporary medical training, has been viewed as both morally and legally unacceptable throughout much of history, and the procurement of bodies for dissection was neither easy or pleasant. In centuries past, anatomists across Europe infamously resorted to robbing graves and cutting down bodies from gallows in order to obtain fresh materials for their research. However, even the latter, legitimate source of bodies was not great enough to meet demand. Relatives of the condemned individual would try to retrieve the body straight away, as it was, very occasionally, possible to revive someone after hanging. There was also a strong belief that a corpse must remain intact if the soul of the deceased was to have a chance of redemption. By the end of the eighteenth century, demand for corpses was steadily increasing, and it was not only the dead that were at risk of being whisked away.

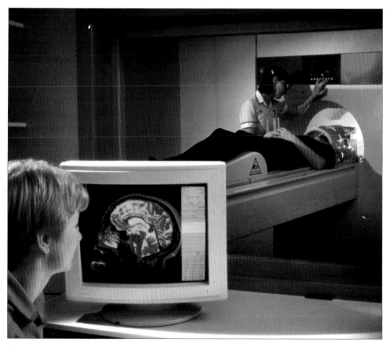

Magnetic resonance imaging (MRI) came to the forefront in the 1990s. Strong magnetic fields cause hydrogen atoms in the body to produce radio waves that form a "sliced" MRI image through the body. This can be used to study tumors in soft tissue, such as the brain.

William Burke and William Hare, Edinburgh-based suppliers of fresh cadavers, maintained their supply by murdering unwitting visitors to their guesthouse, smothering them as they slept.

Finally, to help put an end to these shady practices, the government stepped in, and by the mid-1830s, many U.S. states had passed the Anatomy Act. To destigmatize the practice of dissection, the Act put an end to the use of criminal corpses. Over time, as consent became necessary for dissection, the public gradually lost some of its disgust for and loathing of the practice. Society's attitude toward death also changed with the advent of World War I. People learned to dissassociate the whereabouts or condition of a dead body from the fate of the soul and, slowly, medical schools began to receive a steady supply of legitimately donated corpses.

The late 1970s saw a new invention that allowed for the preservation of cadavers in a new and more versatile way. Invented by German polymer chemist-turned-anatomist Gunther von Hagens, the process of plastination retained the flexibility and natural appearance of anatomical specimens, and at the same time protected them during repeated handling and study. Specimens could be predissected to display underlying structures or, after processing, could be sliced to show a variety of different cross-sectional views. As with cadaver dissection, plastinated specimens afford the student the opportunity to examine important internal structures, with the added benefit of being reuseable and portable. They can be used for learning in many settings and passed around among students during classroom presentations.

At the beginning of the twenty-first century, anatomists are continuing to develop new resources and technology for research and training. Interactive, three-dimensional digital models of the human body on the Internet—created from X-rays, computer tomography, magnetic resonance images, and pictures of actual cross-sections cut through the two cadavers—allow users to simulate injuries and diseases, as well as surgery, on a computer. This technology might completely eliminate the need for cadavers for medical study in medical schools and hospitals, and students can practice surgical procedures over and over again.

DISCOVERING HUMAN ANATOMY

Of course, the human body is a fascinating and complex machine, and an interest in anatomy is certainly not restricted to scientists and medical students. This has been particularly true during the last few decades in which,

In 2001, Gunther von Hagens brought the wonder of anatomy alive for the general public in his Body Works exhibition, which featured hundreds of plastinates in natural, everyday activities and poses, from sitting in a chair to playing basketball.

with the increase in availability and acceptance of alternative therapies, such as acupuncture and yoga, we have been encouraged more and more to think about our health and how our bodies function.

So how much do we actually know about our own body systems, and how can we better understand what the doctor, surgeon, or therapist sees and does? Structured from the head to the toe, and broken down into the head and neck, spine and spinal cord, thorax, arms, abdomen, pelvis, legs, and whole body systems, *The Human Body* will show you what we are really made of. Each section examines the bones, muscles, nerves, soft tissue, and organs, and explains how they work and interact. This book is the beginning of a fascinating journey.

The Skull

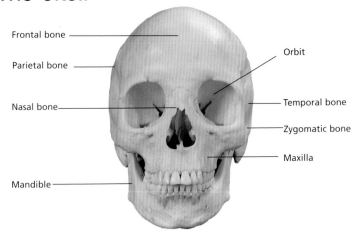

Frontal bone

Parietal bone

Nasal bone

Mandible

Orbit

Temporal bone

Zygomatic bone

Maxilla

The skull is a bony structure located at the top of the backbone. It is the body's natural crash helmet and surrounds the brain and sensory organs (the eyes, ears, and nose), protecting them from damage. Although it appears to be a single bone, the skull is in fact formed from 22 separate parts. The dome-shaped section of the skull is known as the cranial vault and is made up of eight bones, while the remaining 14 are the facial bones. The adult skull is heavy, and it is thought that the four air-filled cavities, or sinuses, in the facial bones may help lighten it. In addition to these cavities, scattered around the base and sides of the skull are small holes through which nerves and blood vessels pass to the brain.

Body system:	skeletal system
Location:	the head
Function:	bony covering that protects the brain and sensory organs (the eyes, ears, and nose) from injury
Components:	22 bones, including cranium and facial bones
Related parts:	vertebral column (backbone), jaw, teeth

Skull Sutures

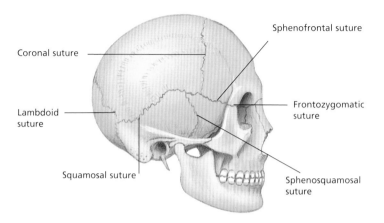

Sphenofrontal suture

Coronal suture

Lambdoid suture

Frontozygomatic suture

Squamosal suture

Sphenosquamosal suture

A suture is a immobile joint found only between the bones of the skull. Before birth, the sutures are flexible and the bones of the skull of a fetus are soft so they can expand as the brain increases in size. This pliable skull also allows the baby to move easily down the birth canal. Even after birth, the bones remain much softer than an adult's bones, and it is possible to feel spongy areas at the front and rear of the cranium. These so-called fontanelles are the fibrous membranes that fill the gaps between the growing cranial bones. By the age of 18 months, a child's brain is almost the same size as an adult's brain and the skull bones have hardened. The sutures are not visible through the skin and the fontanelles have closed.

Body system:	skeletal system
Location:	between the bones of the vault of the skull
Function:	during fetal development, the sutures allow the brain to grow; at birth, these joints allow flexibility in the skull shape
Components:	fibrous cartilage between adjacent bones
Related parts:	skull bones

The Scalp

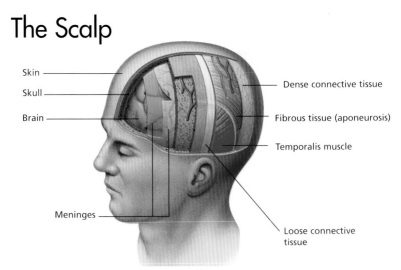

Skin

Skull

Brain

Dense connective tissue

Fibrous tissue (aponeurosis)

Temporalis muscle

Meninges

Loose connective tissue

The scalp is the multilayered protective covering over the top of the head, which stretches from the hairline at the back of the skull to the eyebrows. It is a thick, mobile, protective covering for the skull, and it has five distinct layers, the first three of which are bound tightly together. As well as its protective role, the scalp is a hair-bearing area and provides insulation against sunlight and cold temperatures. In addition, the scalp plays a major role in facial expression as many of the fibers of the scalp muscles are attached to the skin, allowing it to move backward and forward. A rich arterial blood supply is necessary to nourish the many hair follicles, which is why the scalp bleeds so profusely after even a minor injury.

Body system:	integumentary system
Location:	the top of the head
Function:	the scalp forms a tough protective layer around the skull; it is also hair-bearing and protects against sunlight and the cold
Components:	skin, connective tissue, aponeurosis, muscle
Related parts:	skull, hair, ears

The Brain

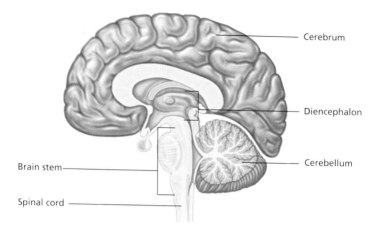

Cerebrum

Diencephalon

Cerebellum

Brain stem

Spinal cord

Weighing about 3 lb (1.5 kg) and with the consistency of cold porridge, the brain is the most complex organ in the body. It not only controls movement and bodily functions but is also the center of intellectual activity and consciousness. There are four main parts to the brain: the cerebrum (consisting of right and left cerebral hemispheres), the cerebellum, the diencephalon (thalamus and hypothalamus), and the brain stem, each of which have vastly different functions. These structures are formed of billions of nerve cells and fibers, all linked to each other and to the spinal cord so that messages in the form of nerve impulses can be relayed throughout the nervous system.

Body system:	central nervous system
Location:	inside the skull
Function:	control center for all the body's functions, as well as for intellectual activity and consciousness
Components:	the cerebrum, cerebellum, diencephalon, and brain stem
Related parts:	spinal cord, peripheral nerves

Inside the Brain

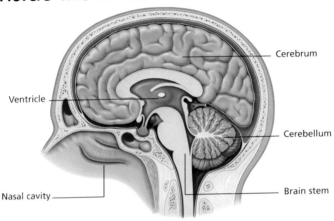

- Cerebrum
- Ventricle
- Cerebellum
- Brain stem
- Nasal cavity

A midline section between the two cerebral hemispheres reveals the main structures in the brain, and their relationship to one another. These structures control a vast number of activities in the body. Some areas of the brain monitor sensory and motor information, while others control speech and sleep. Protected under the two large cerebral hemispheres are the thalamus and hypothalamus, and to the rear is the cerebellum, which is responsible for balance and control of muscle movement. The stalklike brain stem is a vital structure that relays messages between the spinal cord and the brain. It is responsible for subconscious functions, such as breathing and heart rate.

Body system:	central nervous system
Location:	inside the skull
Function:	control center for all the body's functions, as well as for intellectual activity and consciousness
Components:	the cerebrum, cerebellum, diencephalon, and brain stem
Related parts:	spinal cord, peripheral nerves

Gray Matter

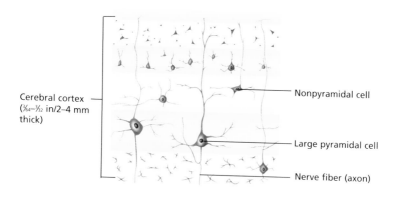

Cerebral cortex
(³⁄₆₄–³⁄₃₂ in/2–4 mm thick)

Nonpyramidal cell

Large pyramidal cell

Nerve fiber (axon)

The brain and spinal cord are composed of two types of tissue—gray matter and white matter. Gray matter is responsible for originating nerve impulses and contains large numbers of nerve-cell bodies. The very outer layer of the cerebral hemispheres (cortex) is gray matter and has six separate layers of cells. These cells differ in shape, but there are two main types: pyramidal and nonpyramidal. Pyramidal cells have axons (nerve fibers) that project out of the cortex and carry information to other regions of the brain. Nonpyramidal cells have smaller bodies and are involved in receiving and analyzing input from other areas of the brain.

Body system:	central nervous system
Location:	the cerebral cortex
Function:	control center for all the body's functions, as well as for intellectual activity and consciousness
Components:	nerve cell bodies and fibers
Related parts:	other parts of the brain and spinal cord

White Matter

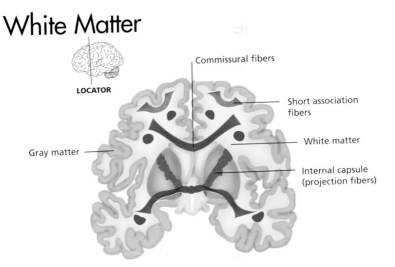

LOCATOR

Commissural fibers

Short association fibers

White matter

Gray matter

Internal capsule (projection fibers)

Underneath the cortex is the white matter of the brain, which makes up the bulk of the cerebral hemispheres. It is composed largely of nerve fibers, rather than cell bodies, and transmits nerve impulses within the brain and to other parts of the body. White matter is arranged into bundles or tracts of three types. Commissural fibers cross between the two hemispheres, connecting corresponding areas on each side. Short- and long-association fibers connect different areas in the same hemisphere, and projection fibers connect the cerebral cortex to the brain stem and the spinal cord. Projection fibers enable the cortex to receive incoming information and to send out instructions controlling movement and bodily functions.

Body system:	central nervous system
Location:	underneath the cerebral cortex
Function:	transmits nerve impulses within the brain and from the brain to other parts of the body
Components:	commissural fibers, association fibers, projection fibers
Related parts:	other structures of the brain and the spinal cord

Arteries of the Brain

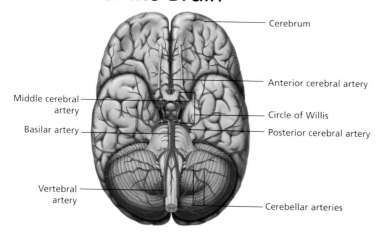

Cerebrum

Anterior cerebral artery

Middle cerebral artery

Circle of Willis

Basilar artery

Posterior cerebral artery

Vertebral artery

Cerebellar arteries

Although it accounts for only 2 percent of our body weight, the brain needs about 15–20 percent of the oxygenated blood pumped out of the heart with each beat to be able to function properly. If the blood supply to the brain is stopped for as little as 10 seconds we lose consciousness, and unless blood flow is quickly restored, it is only a matter of minutes before irreversible brain damage occurs. Blood reaches the brain via two pairs of arteries: the carotid arteries and the vertebral arteries, both of which originate in the lower neck and chest. These two sources of blood to the brain are linked by other arteries to form a circuit at the base of the brain known as the "circle of Willis."

Body system:	cardiovascular system
Location:	surrounding the brain
Function:	to supply blood rich in oxygen and nutrients to the tissues of the brain
Components:	cerebral arteries, basilar artery, circle of Willis, vertebral artery, cerebellar arteries
Related parts:	brain tissue, heart

Veins of the Brain

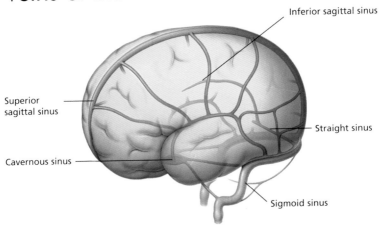

Inferior sagittal sinus

Superior sagittal sinus

Straight sinus

Cavernous sinus

Sigmoid sinus

The veins of the brain can be divided into deep and superficial groups. These veins, unlike other veins in the body, do not have one-way valves and rely on gravity to return blood to the heart. The veins drain deoxygenated blood from the surface of the brain and from the deep structures into a complex system of venous sinuses formed between layers of dura mater, the tough membranous covering of the brain. Most superficial veins drain into the superior sagittal sinus, while the very deep veins drain into the straight sinus via the great cerebral vein—the vein of Galen. These two sinuses converge and blood leaves the brain in the internal jugular vein and passes downward toward the heart.

Body system:	cardiovascular system
Location:	surrounding the brain
Function:	to drain deoxygenated blood from the brain tissue; sinuses also drain cerebrospinal fluid
Components:	various deep and superficial veins and sinuses
Related parts:	brain tissue

Meninges

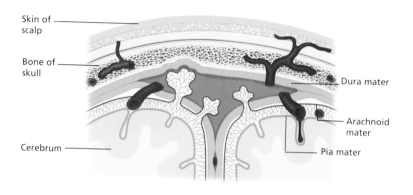

Skin of scalp

Bone of skull

Cerebrum

Dura mater

Arachnoid mater

Pia mater

Between the skull and the brain, and around the spinal cord, there is added protection in the form of three tough membranes called meninges. The outer membrane—the dura mater—is a thick fibrous tissue that lines the inner surface of the skull and contains blood vessels that supply the cranial bones. The arachnoid mater is an impermeable membrane that follows the contours of the dura mater and is separated from it by a tiny gap called the subdural space. The third membrane, the pia mater, covers the brain and spinal cord. The space between the arachnoid and pia maters (the subarachnoid space) is filled with cerebrospinal fluid, which bathes the brain and spinal cord and provides a buffer against injury.

Body system:	central nervous system
Location:	surrounding the brain and spinal cord
Function:	protect the brain and spinal cord and provide a space within which cerebrospinal fluid flows
Components:	arachnoid mater, dura mater, and pia mater
Related parts:	brain and spinal cord

Dural Venous Sinuses

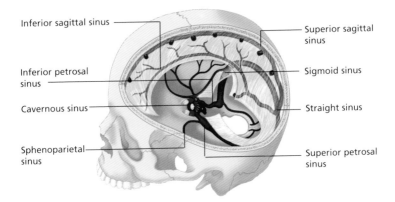

Inferior sagittal sinus

Superior sagittal sinus

Inferior petrosal sinus

Sigmoid sinus

Cavernous sinus

Straight sinus

Sphenoparietal sinus

Superior petrosal sinus

Between the folds of the dura mater (the outermost of the three membranes that surround the brain and spinal cord) are 15 dural venous sinuses, which play a role in the circulation and drainage of the blood and fluids that protect and bathe the brain. Venous sinuses are delicate blood-filled cavities lined with endothelium, which rely on the surrounding tissues for support because they have no muscular tissue within their walls. There are two sets of dural sinuses, those in the upper part of the skull and those on the floor of the skull. They receive blood draining from the brain, the scalp, and the red bone marrow of the skull. The dural venous sinuses are also crucial to the reabsorption of cerebrospinal fluid.

Body system:	cardiovascular system
Location:	between the folds of the dura mater, the outermost of the three membranes that surround the brain
Function:	receive blood draining from the brain; also play a part in the drainage of cerebrospinal fluid
Components:	two sets of sinuses
Related parts:	deep and superficial veins supplying the brain, dura mater

Ventricles

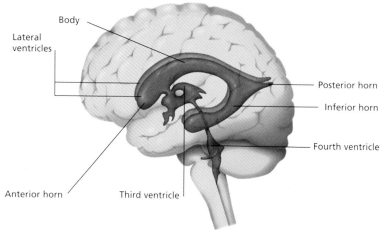

Body

Lateral ventricles

Posterior horn

Inferior horn

Fourth ventricle

Anterior horn

Third ventricle

The brain "floats" in a protective layer of cerebrospinal fluid (CSF), which is produced by specialized cells in four cavities in the brain known as ventricles. Within each ventricle is a network of blood vessels known as the choroid plexus, and it is here that CSF is manufactured. The ventricles are connected to each other, and to the spinal cord, by narrow channels. Three of the ventricles—the paired lateral ventricles and the third ventricle—are located toward the front of the brain. The fourth ventricle is situated at the back of the brain in front of the cerebellum and is connected to the third ventricle by the "cerebral aqueduct." The two lateral ventricles consist of a "body" and three "horns."

Body system:	central nervous system
Location:	inside the brain
Function:	produce cerebrospinal fluid, which acts as a buffer and protects the brain
Components:	two lateral ventricles, third and fourth ventricles
Related parts:	brain and spinal cord

Circulation of CSF

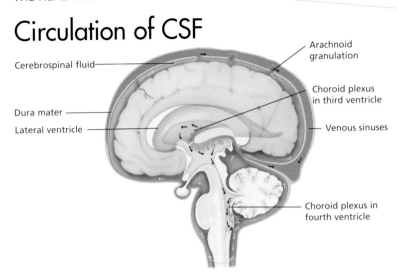

Cerebrospinal fluid

Arachnoid granulation

Choroid plexus in third ventricle

Dura mater

Lateral ventricle

Venous sinuses

Choroid plexus in fourth ventricle

Cerebrospinal fluid (CSF) is a clear watery substance produced by networks of capillary blood vessels, known as the choroid plexuses, located in the lateral, third, and fourth ventricles. CSF flows between the ventricles and around the brain and spinal cord in the subarachnoid space. As the fluid is produced continuously, there must be constant absorption to prevent a dangerous buildup of fluid within the skull (normally there is between 3–5 oz (80–150 ml) of fluid circulating in the nervous system). This is achieved by the passage of of the CSF into the venous sinuses of the brain through fingerlike projections, known as arachnoid granulations, and from there into the general venous circulation.

Body system:	central nervous system
Location:	surrounding the brain and spinal cord
Function:	cerebrospinal fluid protects the brain and spinal cord from injury as it cushions the organs
Components:	choroid plexuses, ventricles, subarachnoid space, venous sinuses
Related parts:	brain, spinal cord

The Cerebrum

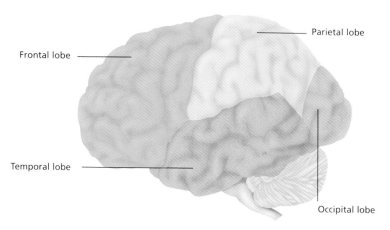

Parietal lobe

Frontal lobe

Temporal lobe

Occipital lobe

The cerebrum consists of the right and left cerebral hemispheres, which between them form the main bulk of the brain and constitute seven-eighths of its total weight. Each hemisphere is further divided into four lobes named after the bones of the skull that lie over them: the frontal lobe, parietal lobe, occipital lobe, and temporal lobe. The two hemispheres are separated by the longitudinal fissure, which runs between them. Sulci (shallow grooves) and fissures (deep grooves) also separate the four lobes. Two distinct types of nervous tissue make up the cerebrum—"gray matter" forms the cortex and underneath this outer layer is the brain's "white matter," as well as islands of gray matter.

Body system:	central nervous system
Location:	within the vault of the skull
Function:	motor and sensory functions and higher mental processes
Components:	left and right cerebral hemispheres
Related parts:	cerebellum, brainstem, diencephalon, spinal cord, peripheral nerves

Gyri and Sulci

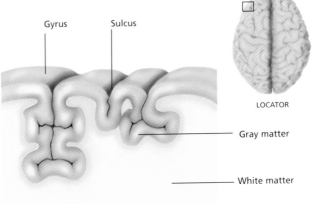

Gyrus   Sulcus

LOCATOR

Gray matter

White matter

Almost the entire surface of the cerebral hemispheres is a convoluted arrangement of ridges and furrows. As a baby's brain grows rapidly before birth, the cerebral cortex folds in on itself, producing the characteristic appearance of a walnut. This complicated folding allows a larger surface area of cerebral cortex to be contained within the confined space of the skull. The folds are known as gyri and the shallow grooves between them are the sulci. Certain sulci are found in the same position in all human brains, and these are used as landmarks to divide the cortex into four lobes. Deeper grooves are known as fissures; the largest of these is the longitudinal fissure, which separates the two cerebral hemispheres.

Body system:	central nervous system
Location:	cerebrum
Function:	increase the surface area of the cortex; some sulci are boundaries between lobes
Components:	gray matter, white matter
Related parts:	cerebellum, diencephalon, brainstem, meninges, skull

Functions of the Cortex

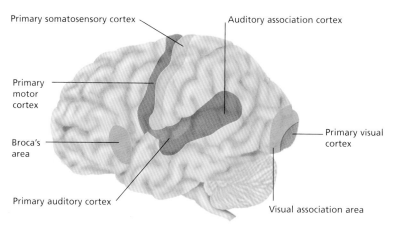

Primary somatosensory cortex

Auditory association cortex

Primary motor cortex

Broca's area

Primary auditory cortex

Primary visual cortex

Visual association area

The surface of the cerebrum, the cortex, is the area that processes most of the brain's information. Different regions of the cortex have distinct and highly specialized functions. Motor areas initiate and control voluntary movement, including complex sequences of finely controlled movement; the primary motor cortex controls voluntary movement of the opposite side of the body. Sensory areas receive and integrate information from around the body, such as pain impulses, temperature, touch, and the position of joints and muscles (proprioception). Association areas are involved with more complex brain functions, such as learning, memory, language, judgement, emotion, and personality.

Body system:	central nervous system
Location:	cerebrum
Function:	motor and sensory functions and higher mental processes, such as learning and memory
Components:	gray matter
Related parts:	cerebellum, diencephalon, brainstem, meninges, skull

Thalamus

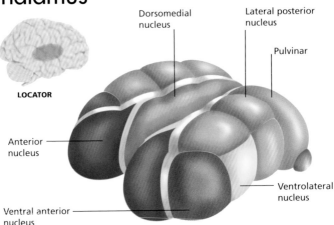

Dorsomedial
nucleus

Lateral posterior
nucleus

Pulvinar

LOCATOR

Anterior
nucleus

Ventrolateral
nucleus

Ventral anterior
nucleus

The thalamus consists of a pair of egg-shaped masses of mainly gray matter located deep in the brain. The two structures lie on either side of the fluid-filled third ventricle (cavity) and are connected by a bridge of gray matter. The thalamus is an important relay station for impulses traveling from the spinal cord and other parts of the brain to the cerebral cortex. Collections of nerve cells called nuclei in the thalamus have specific functions. Some relay sensory impulses, such as touch and pain, to the appropriate area of the cerebral cortex, while others direct information concerning the motor (movement) system to the motor regions of the frontal cortex. The thalamus is also involved in autonomic (unconscious) functions.

Body system:	central nervous system
Location:	between the cerebral hemispheres, either side of the third ventricle
Function:	relay station for sensory and motor information; also associated with some autonomic functions
Components:	gray matter, nuclei (clusters of nerve cells)
Related parts:	cerebrum, cerebellum, brainstem, spinal cord

Hypothalamus

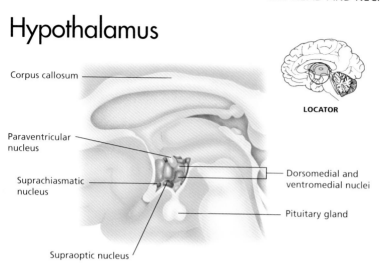

Corpus callosum

LOCATOR

Paraventricular
nucleus

Suprachiasmatic
nucleus

Dorsomedial and
ventromedial nuclei

Pituitary gland

Supraoptic nucleus

The hypothalamus is the size of a thumbnail, but despite its small size, it has a range of important functions. It is composed of a number of nuclei (clusters of nerve cells), which are involved in the control of many autonomic (unconscious) functions, such as interpreting smell, regulating the heart rate, and controlling the passage of food through the gut. The hypothalamus is the main link between the nervous system and the endocrine system and secretes chemicals that influence the production of hormones. Hunger and thirst centers are located within the hypothalamus, as is a thermostat that regulates body temperature. Emotions, such as fear, rage, and aggression, are also associated with the hypothalamus.

Body system:	central nervous system
Location:	under the thalamus in the center of the brain
Function:	controls many autonomic (unconscious) functions such as the rate of the heartbeat, fluid balance, and body temperature
Components:	nuclei (clusters of nerve cells)
Related parts:	pituitary gland, autonomic nervous system,

Limbic System

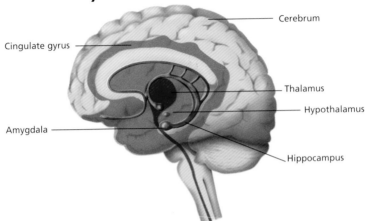

Cerebrum

Cingulate gyrus

Thalamus

Hypothalamus

Amygdala

Hippocampus

The limbic system is a part of the brain associated with the perception of emotions and the body's response to them. There is also a close link between smell and the limbic system, particularly where a particular odor is linked to a memory or an emotion. The system consists of five interconnected structures that surround the upper part of the brain stem. The amygdala appears to be linked to feelings of fear and aggression, whereas the hippocampus is believed to play a role in learning and memory. The hypothalamus regulates the body's internal environment and the anterior hypothalamic nuclei control instinctive drives. The cingulate gyrus connects the limbic system to the cerebral cortex.

Body system:	central nervous system
Location:	deep within the cerebrum of the brain
Function:	relays information from senses to cortex; responsible for learning, memory, and emotion
Components:	amygdala, hippocampus, hypothalamus, cingulate gyrus, anterior thalamic nuclei
Related parts:	other areas of the brain

Basal Ganglia

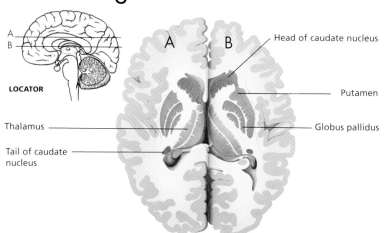

LOCATOR

A

B

Head of caudate nucleus

Putamen

Thalamus

Globus pallidus

Tail of caudate nucleus

The basal ganglia (sometimes known as basal nuclei) are paired areas of gray matter in each of the cerebral hemispheres. They are primarily responsible for the control of different kinds of movement. There are a number of component parts to the basal ganglia that are anatomically and functionally closely related to each other. The largest of these is the corpus striatum, which consists of the caudate nucleus and the lentiform nucleus, which in turn is divided into the putamen and the globus pallidus. It is thought that the basal ganglia function in association with the cerebral cortex to produce appropriate movement and inhibit unwanted or inappropriate movement.

Body system:	central nervous system
Location:	in each cerebral hemisphere
Function:	involved in the control of movement
Components:	caudate nucleus, lentiform nucleus, putamen, globus pallidus
Related parts:	other parts of the brain, peripheral nervous system

Structure and Role of Basal Ganglia

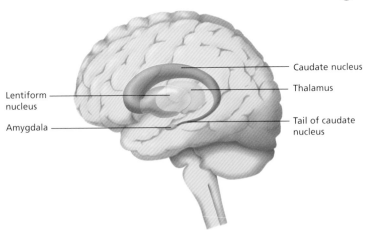

Caudate nucleus

Thalamus

Lentiform nucleus

Amygdala

Tail of caudate nucleus

When seen in a three-dimensional view, the size and shape of the basal ganglia, together with their position in the brain, can be seen more easily. The functions of the basal ganglia have proved difficult to understand because they lie deep within the brain and are relatively inaccessible. Much of what is known about their role has been surmised from the symptoms of people who have disorders of the basal ganglia, such as in Parkinson's disease. It is thought that the caudate nucleus and the putamen control subconscious movements of the skeletal muscles, such as swinging the arms when walking. The globus pallidus is involved with regulating muscle tone during specific movements.

Body system:	central nervous system
Location:	in each cerebral hemisphere
Function:	help to produce appropriate movement
Components:	caudate nucleus, lentiform nucleus, putamen, globus pallidus
Related parts:	other parts of the brain, peripheral nervous system

Cerebellum

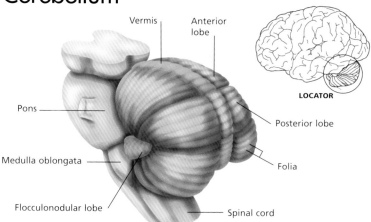

Vermis

Anterior lobe

Pons

Medulla oblongata

Flocculonodular lobe

LOCATOR

Posterior lobe

Folia

Spinal cord

The cerebellum is the second-largest part of the brain and is located toward the rear, tucked away under the cerebral hemispheres. It is a butterfly-shaped structure that has two hemispheres joined to each other by the wormlike part of the cerebellum known as the vermis. The cerebellum has a very distinctive appearance—in contrast to the large folds of the cerebral hemispheres, its surface is made up of numerous fine folds (folia). Between the folia of the cerebellar surface lie deep fissures, which divide it into three lobes: the anterior, posterior, and flocculonodular lobes. The cerebellum is responsible for the coordination of movement and the maintenance of balance and posture.

Body system:	central nervous system
Location:	to the rear of the brain, just under the cerebral hemispheres
Function:	coordination of movement and maintenance of balance and posture
Components:	two hemispheres, each consisting of three lobes
Related parts:	other parts of the brain, peripheral nerves

Cerebellar Peduncles

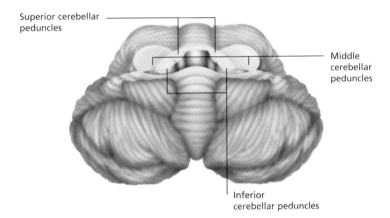

Superior cerebellar peduncles

Middle cerebellar peduncles

Inferior cerebellar peduncles

The cerebellum is attached to the brain stem, and to the rest of the brain, by three pairs of nerve fiber tracts, which make up the cerebellar peduncles. The superior cerebellar peduncles connect the cerebellum to the part of the brain stem called the midbrain; the middle cerebellar peduncles provide a link to the pons; and the inferior cerebellar peduncles form a connection with the medulla. There are no direct connections between the cerebellum and the cerebral cortex, so all information to and from the cerebellum goes through the peduncles. Unlike the cerebral cortex, where each hemisphere controls the opposite side of the body, each half of the cerebellum controls the same side of the body.

Body system :	central nervous system
Location:	between the cerebellum and the brainstem
Function:	provide a link between the cerebellum and the rest of the brain and a pathway for nerve impulses from the cerebral cortex
Components:	superior, inferior, and middle peduncles
Related parts:	right and left hemispheres of the cerebrum, brain stem

Brain Stem

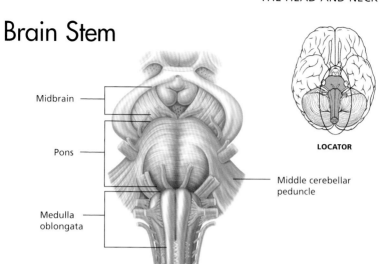

Midbrain

Pons

Medulla
oblongata

LOCATOR

Middle cerebellar
peduncle

The brain stem is situated at the junction of the brain and the spinal cord and provides pathways for all nerve impulses traveling between the upper parts of the brain and the rest of the body. There are three parts to the brain stem: the midbrain, the pons, and the medulla oblongata. Together, they are responsible for many of the automatic functions that are necessary for survival, such as maintaining the normal rhythm of breathing and heart beat. Responses to visual and auditory stimuli that influence head movement are also controlled here. It is in the medulla that nerve tracts descending from the cerebral cortex cross over, with the result that each cerebral hemisphere controls movement on the opposite side of the body.

Body system:	central nervous system
Location:	at the junction of the spinal cord and the brain
Function:	provides nerve pathways for ascending and descending impulses; regulates many vital body functions, such as breathing
Components:	midbrain, pons, and medulla
Related parts:	spinal cord, other parts of the brain

Structure of the Brain Stem

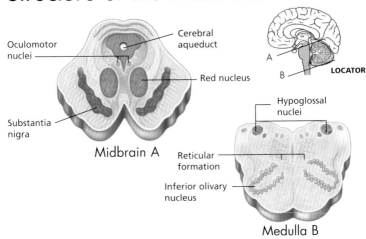

Oculomotor nuclei

Cerebral aqueduct

Red nucleus

Substantia nigra

Midbrain A

LOCATOR

A

B

Hypoglossal nuclei

Reticular formation

Inferior olivary nucleus

Medulla B

The brain stem contains many areas of nerve tissue that have several functions vital to life. These two cross-sections through the brain stem reveal its internal structure and the arrangement of white and gray matter, which varies according to the level of the section. In the midbrain the cerebral aqueduct is shown, a channel that connects the third and fourth ventricles (fluid-filled cavities) in the brain. In each cerebellar peduncle are two structures; the substantia nigra, damage to which is linked with Parkinson's disease, and the red nuclei, which are involved with control of movement. In the medulla, the complex network of neurones called the reticular formation controls respiration and circulation.

Body system:	central nervous system
Location:	at the junction of the spinal cord and brain
Function:	provides nerve pathways for ascending and descending impulses; regulates many vital body functions, such as breathing
Components:	midbrain, pons, and medulla
Related parts:	spinal cord, other parts of the brain, cranial nerves

Cranial Nerves

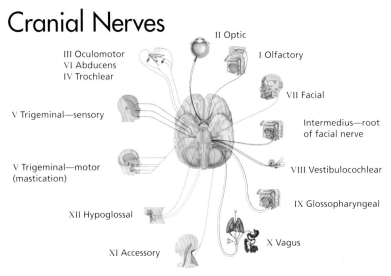

II Optic

III Oculomotor
VI Abducens
IV Trochlear

I Olfactory

VII Facial

V Trigeminal—sensory

Intermedius—root
of facial nerve

V Trigeminal—motor
(mastication)

VIII Vestibulocochlear

IX Glossopharyngeal

XII Hypoglossal

X Vagus

XI Accessory

Nerves are the routes by which information passes between the brain and the rest of the body. Most of these nerves emerge from the spinal cord, passing out through tiny openings in the bony spinal column; however, the cranial nerves emerge directly from the brain. There are 12 pairs of cranial nerves that leave the brain to supply structures mainly in the head and neck. They are made up of different nerve fibers: sensory fibers bring information about pain, temperature, and touch, as well as the senses of taste, vision, and hearing; motor fibers send instructions to the head and neck muscles, allowing movement; autonomic nerve fibers allow subconscious control of internal structures, such as the salivary glands.

Body system:	central nervous system
Location:	emerging from the brain stem and the forebrain
Function:	many sensory (smell, hearing, taste), motor (facial expression) and autonomic functions
Components:	cranial nerves I– XII
Related parts:	brain, eyes, ears, nose, muscles

Olfactory Nerves

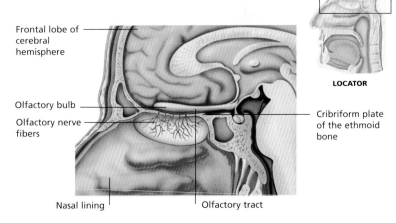

Frontal lobe of cerebral hemisphere

Olfactory bulb

Olfactory nerve fibers

Nasal lining

Olfactory tract

LOCATOR

Cribriform plate of the ethmoid bone

The olfactory nerves are tiny sensory nerves of smell that run from the lining of the nose cavity (nasal mucosa) to the olfactory bulb in the brain. Specialized nerve cells in the nasal mucosa called receptor cells detect odorous substances present in the air in the form of minute droplets and pass "odor signals" to their long axons. These axons are grouped together to form about 20 bundles, which pass up through a thin plate of bone at the base of the skull, the cribriform plate of the ethmoid bone, to reach the olfactory bulbs. Odor signals reach the olfactory bulb and pass across to mitral nerves. These carry the information to the olfactory centers in the brain, where the smell is analyzed and responded to.

Body system :	central nervous system
Location:	run from the nasal mucosa to the olfactory bulb
Function:	transmit information about smell from receptor cells in the nasal mucosa to the brain
Components:	receptor cells, axons
Related parts:	nose, olfactory bulb, olfactory center in brain

Optic Nerve

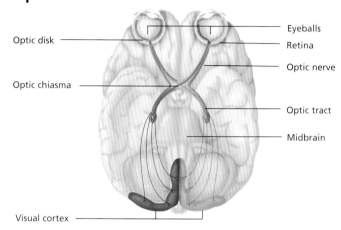

Optic disk

Optic chiasma

Visual cortex

Eyeballs

Retina

Optic nerve

Optic tract

Midbrain

The optic nerve carries information from the retina (the light-sensitive cells at the back of the eye) to the visual cortex of the brain for processing. Unlike many of the other cranial nerves, the optic nerve is sensory only, which means that it only takes information to the brain, not from it. The optic nerve is formed from the axons, or nerve fibers, of the retinal cells. These join together to form the nerve, which leaves the back of the eyeball at a point called the optic disk. As the nerves enter the skull through the optic canal, they converge at a point called the optic chiasma, where some of the nerve fibers cross over to the other side. The nerve fibers continue as optic tracts until they reach the thalamus.

Body system:	central nervous system
Location:	run from the back of the eye to the visual cortex
Function:	transmit visual information from the retina (the light-sensitive cells at the back of the eye) to the brain for interpretation
Components:	nerve fibers, optic chiasma, optic tracts
Related parts:	eye and other parts of the brain

Oculomotor, Trochlear, and Abducent Nerves

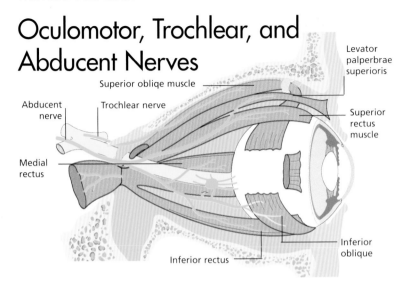

Levator palperbrae superioris

Superior obliqe muscle

Abducent nerve

Trochlear nerve

Superior rectus muscle

Medial rectus

Inferior oblique

Inferior rectus

Between them, the oculomotor, trochlear, and abducent nerves supply the six muscles responsible for eye movement. The nerves do not carry any information relating to sight. All three nerves carry both motor fibers that take "instructions" to the muscles and sensory fibers, which relay sensory information concerning the position of the muscles back to the brain. In addition, the oculomotor nerve contains some fibers from the autonomic nervous system that constrict the pupil and alter the lens shape. The small trochlear nerve supplies the superior oblique muscle and the abducent nerve supplies the lateral rectus muscle. The oculomotor nerve supplies the remaining four muscles.

Body system:	central nervous system
Location:	originate in the brain and supply the muscles of the eye
Function:	movement of the eyelid and eyeball, constriction of pupil and alteration of the shape of the lens
Components:	sensory, motor, and autonomic nerve fibers
Related parts:	muscles of movement of the eye, brain

Trigeminal Nerve

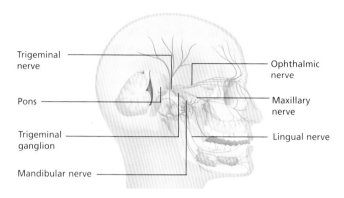

Trigeminal nerve

Pons

Trigeminal ganglion

Mandibular nerve

Ophthalmic nerve

Maxillary nerve

Lingual nerve

The trigeminal nerve is the largest of the cranial nerves and has three branches: the ophthalmic, maxillary, and mandibular branches. Sensory information, such as pain and touch from the face, scalp, cornea, and the nasal and oral cavities, travels via the trigeminal nerve to the pons (part of the brain stem) to be processed. On the way, information passes through the trigeminal ganglion, an expansion of the nerve. The trigeminal nerve also has a role to play in the control of some important muscles, mainly those involved with mastication (chewing), such as the masseter muscle and the temporalis. Only the lower branch of the trigeminal nerve, the mandibular nerve carries motor fibers that supply these muscles.

Body system:	central nervous system
Location:	originates in the brain and supplies the face, scalp, and oral and nasal cavities
Function:	sensory and motor functions; relays sensory information to the brain and is involved in control of some facial muscles
Components:	ophthalmic, maxillary, and mandibular branches
Related parts:	brain, eyes, ears, nose, muscles of mastication

Facial Nerve

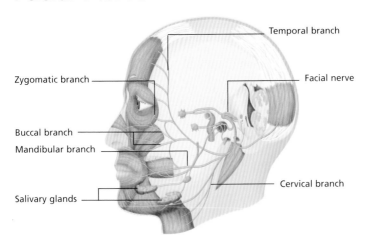

Temporal branch

Zygomatic branch

Facial nerve

Buccal branch

Mandibular branch

Cervical branch

Salivary glands

As the facial nerve emerges from the skull, it divides into six main branches, each of which supplies a different area. There are three types of fibers in the facial nerve—motor, sensory, and autonomic. The motor fibers supply the muscles of facial expression, such as those that enable smiling and frowning, as well as scalp muscles. Sensory fibers in the facial nerve transmit information about taste from the tongue back to the brain for interpretation. The nerve also contains parasympathetic fibers, which are concerned with the unconscious regulation of the body's internal environment. These supply the lacrimal (tear) duct and salivary glands and help to regulate the production of tears and saliva.

Body system:	central nervous system
Location:	originates in the brain and supplies the face and other structures
Function:	supplies facial and scalp muscles, transmits sensory information from the tongue to the brain, regulates salivary and tear production
Components:	posterior auricular, temporal, zygomatic, buccal, mandibular, and cervical branches
Related parts:	tongue, facial muscles, salivary glands, lacrimal ducts, brain

Vestibulocochlear Nerve

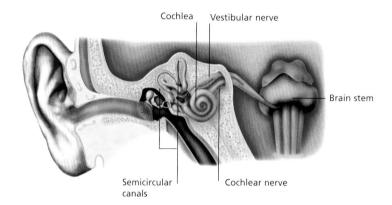

Cochlea | Vestibular nerve

Brain stem

Semicircular canals

Cochlear nerve

The vestibulocochlear nerve is a sensory nerve that carries information about balance and hearing from the inner ear to the auditory cortex. The nerve consists of two separate parts—the vestibular and cochlear nerves—which correspond to these two areas of the inner ear. The vestibular nerve carries information about the position and movement of the head from movement-sensitive hair cells in the semicircular canals and vestibule of the inner ear. The cochlear nerve transmits information about sound from hearing receptors in the organ of Corti within the cochlea of the inner ear. As they leave the inner ear, the two nerves join together to form the vestibulocochlear nerve and continue to the brain stem.

Body system:	central nervous system
Location:	runs from the inner ear to the brain stem
Function:	transmits information about sound, position, and balance to the brain for interpretation
Components:	vestibular nerve, cochlear nerve
Related parts:	inner ear, brain

Auditory Pathway

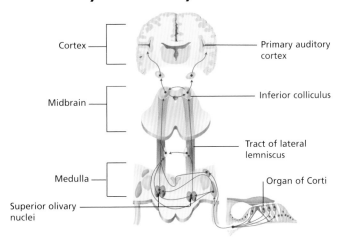

Cortex — Primary auditory cortex

Midbrain — Inferior colliculus

Tract of lateral lemniscus

Medulla — Organ of Corti

Superior olivary nuclei

The perception of sound involves the passage of information along a complex auditory pathway, which runs from the inner ear to the brain. The pathway begins in the organ of Corti in the cochlea, where sounds are converted into electrical impulses. These impulses pass along the cochlear nerve to the superior olivary nuclei in the medulla, and from there up the tract of the lateral lemniscus to the inferior colliculus in the midbrain. Finally, the impulses continue their journey upward via the thalamus to the primary auditory cortex in the temporal lobe of the cerebellum. The area around the auditory cortex is known as Wernicke's area and it is here that information about sound is analyzed and interpreted.

Body system:	special senses
Location:	pathway runs from the inner ear to the cortex of the brain
Function:	provides nerve pathways for ascending information relating to sound
Components:	organ of Corti, cochlear nerve, superior olivary nuclei, tract of lateral lemniscus, inferior colliculus, thalamus, cortex
Related parts:	ear, cortex of the brain

Vagus Nerve

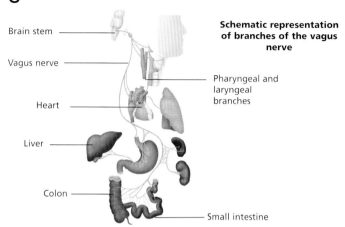

Brain stem

Vagus nerve

Heart

Liver

Colon

Schematic representation of branches of the vagus nerve

Pharyngeal and laryngeal branches

Small intestine

The vagus nerve is the largest of the cranial nerves, extending from the head down to the abdomen. Its role in monitoring and controlling breathing and digestion makes it vital to life. The nerve contains sensory fibers that provide sensation from the lower pharynx (throat), larynx (voicebox), and the organs of the chest and abdomen. Information about taste from the rear of the tongue and throat also travels via the vagus nerve. Motor fibers supply the muscles of the soft palate (the roof of the mouth), the pharynx, and the larynx. The vagus also provides the parasympathetic nerve supply to the internal organs of the chest and abdomen and is crucial in maintaining normal organ activity.

Body system:	central nervous system
Location:	extends from the brain to the abdomen
Function:	carries sensory information from the tongue, throat, chest, and abdominal organs; supplies muscles in the throat and larynx; provides parasympathetic supply to abdominal and chest organs
Components:	motor, sensory and parasympathetic nerve fibers, ganglia
Related parts:	mouth, pharynx, larynx, chest, and abdominal organs, brain

Glossopharyngeal Nerve

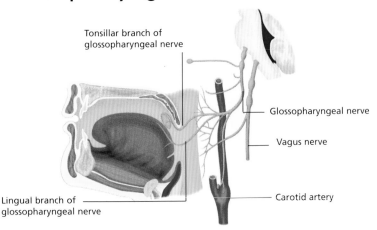

Tonsillar branch of
glossopharyngeal nerve

Glossopharyngeal nerve

Vagus nerve

Carotid artery

Lingual branch of
glossopharyngeal nerve

The glossopharyngeal nerve serves mainly the tongue and the pharynx (throat). It carries both sensory and motor fibers, and also fibers of the parasympathetic branch of the autonomic nervous system. Sensory information taken back to the brain by the glossopharyngeal nerve includes taste and sensation from the back third of the tongue and sensation from the lining of the pharynx (throat). Information about blood oxygen levels is also transmitted via the glossopharyngeal nerve from the carotid body (tissue within the carotid artery in the neck). The motor fibers of the nerve carry impulses to the stylopharyngeus muscle—one of the longitudinal muscles in the pharynx used in swallowing and speaking.

Body system :	central nervous system
Location:	runs from the brain stem to the back of the throat and tongue
Function:	transmits information about taste, sensation, and blood oxygen levels to the brain; supplies a muscle used in swallowing
Components:	sensory, motor, and parasympathetic nerve fibers
Related parts:	tongue, throat, brain

Accessory Nerve

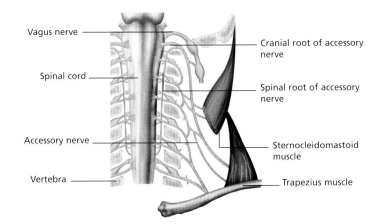

Vagus nerve

Spinal cord

Accessory nerve

Vertebra

Cranial root of accessory nerve

Spinal root of accessory nerve

Sternocleidomastoid muscle

Trapezius muscle

The accessory nerve is unique among the cranial nerves in that it has a spinal root (arising from the spinal cord), as well as a cranial root (arising from the brain stem). The nerve exits the skull through the jugular foramen, at which point the two roots separate again to complete their different functions. Fibers from the cranial root join the large vagus nerve and go on to supply the muscles of the soft palate, pharynx, larynx, and esophagus. Fibers from the spinal root run down as the accessory nerve, lying alongside the internal carotid artery, to reach the sternocleidomastoid muscle and the large trapezius muscle at the base of the neck. These two muscles are responsible for moving the head and neck.

Body system:	central nervous system
Location:	run from the brain stem and spinal cord to the neck
Function:	motor in function—supplies larynx, pharynx, and esophagus, and the trapezius and sternocleidomastoid muscles
Components:	cranial root, spinal root
Related parts:	larynx, pharynx, and esophagus, and trapezius and sternocleidomastoid muscles, brain

Hypoglossal Nerve

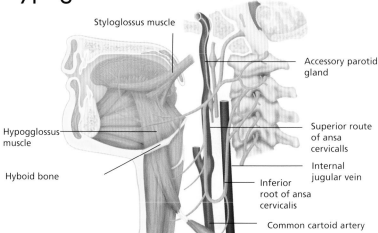

Styloglossus muscle

Accessory parotid gland

Hypogglossus muscle

Hyboid bone

Superior route of ansa cervicalls

Internal jugular vein

Inferior root of ansa cervicalis

Common cartoid artery

The twelfth cranial nerve, whose name literally means "under the tongue," supplies the muscles of the tongue—the styloglossus, hyoglossus, and genioglossus muscles. The hypoglossus nerve has an important role in the actions of chewing, swallowing, and speaking. This cranial nerve is also joined by fibers from the first cervical nerve, which go on to supply other structures. These include muscles attached to the hyoid bone in the neck, which provides a base for tongue movements. In addition a branch of the hypoglossal nerve, called the meningeal branch, carries sensory information from the dura mater (tough protective membrane) surrounding the rear part of the brain.

Body system:	central nervous system
Location:	originates in the brain stem and passes to structures in the neck
Function:	supplies the muscles of the tongue and some in the neck; branches also supply the dura mater at the rear of the brain
Components:	motor nerve fibers
Related parts:	tongue, dura mater, brain

Facial Muscles

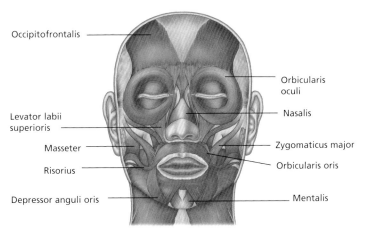

Occipitofrontalis

Orbicularis oculi

Levator labii superioris

Nasalis

Masseter

Zygomaticus major

Risorius

Orbicularis oris

Depressor anguli oris

Mentalis

Just under the skin of the scalp and face lie a group of very thin muscles, which are collectively known as the facial muscles. These muscles enable facial expressions, providing a means of nonverbal communication, and also form sphincters that open and close the eyes and mouth. The majority of facial muscles are attached to the skull bone at one end and to the deep layer of skin at the other. Contraction and relaxation of the muscles alters facial expression and enables a person to articulate speech. A number of small muscles called "dilators" open the mouth. These radiate out from the corners of the lips, and open and close the mouth and pull the lips upward, downward and sideways.

Body system:	musculoskeletal system
Location:	attached to the skull and the deep layer of facial skin
Function:	allow facial expression and the ability to open and close the mouth and eyes, help to articulate speech
Components:	occipitofrontalis, orbicularis oculi, nasalis, zygomaticus major, orbicularis oris, mentalis, depressor anguli oris, masseter, etc.
Related parts:	scalp, skull, face

51

Muscles of Mastication

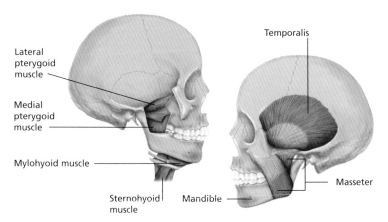

Lateral pterygoid muscle

Medial pterygoid muscle

Mylohyoid muscle

Sternohyoid muscle

Temporalis

Masseter

Mandible

The muscles of mastication (chewing) are those that move the mandible (jawbone) up and down and forward and backward, resulting in the opening and closing of the mouth. The action is also used in other activities, such as speaking and yawning. The temporalis is a fan-shaped muscle connecting the frontal bone of the skull to the mandible; it lifts and retracts the mandible. The masseter is a powerful, thick muscle stretching from the zygomatic arch to under the mandible, and it is the principal muscle used in closing the jaw. Smaller muscles, such as the pterygoid muscles, provide a grinding action. The sternohyoid muscle in the neck assists in swallowing saliva, food, and liquids and plays a role in speaking.

Body system :	musculoskeletal system
Location:	head and neck
Function:	these muscles enable mastication (chewing); they also play a role in speech and swallowing
Components:	temporalis, masseter, pterygoids, sternohyoid muscle
Related parts:	skull, jawbone, hyoid bone

Platysma

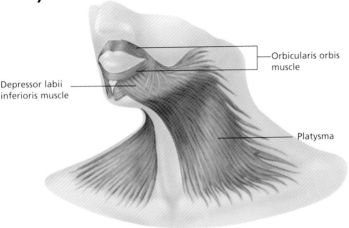

Orbicularis orbis muscle

Depressor labii inferioris muscle

Platysma

The platysma is a large, flat muscle close to the surface of the skin in the neck. Although not strictly a muscle of the head, the platysma plays an important role in facial expression. This thin sheet of muscle extends from below the clavicle (collar bone) up to the mandible (jawbone). It covers the front of the neck, where it tightens the skin, and connects to the muscle and skin at the corners of the mouth. The platysma's role in altering facial expression is to pull the neck skin out and lower the mandible. This pulls the mouth down, as in an expression of disgust. The muscle also assists in the movement of the lower lip. In males, the platysma is the muscle that is tensed while shaving under the chin.

Body system:	musculoskeletal system
Location:	neck
Function:	plays a role in facial expression
Components:	skeletal muscle fibers
Related parts:	facial muscles, jawbone

Opening and Closing the Eye

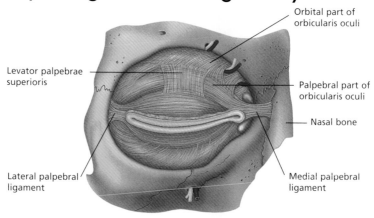

Orbital part of
orbicularis oculi

Levator palpebrae
superioris

Palpebral part of
orbicularis oculi

Nasal bone

Lateral palpebral
ligament

Medial palpebral
ligament

The orbicularis oculi is the muscle responsible for closing the eye. This flat sphincter muscle lines the rim of the eye socket, and various sections of it can be manipulated individually. Part of the orbicularis oculi (the palpebral part) lies in the eyelid, and this section of the muscle closes the eye lightly as in sleeping or blinking. This action also aids the flow of tears across the surface of the eye, keeping it lubricated. A larger part of the orbicularis oculi (orbital part) covers the front of the eye socket and closes the eyes tightly to protect against a blow or bright light. The second orbital muscle is the levator palpebrae superioris. This small muscle pulls on the upper eyelid to open the eye.

Body system:	musculoskeletal system
Location:	around the eye socket
Function:	opening and closing the eye, allowing blinking and sleeping; the action also spreads lubrication across the front of the eye
Components:	orbicularis oculi, levator palpebrae superioris
Related parts:	eyeball, bones of the skull

Arteries of the Head and Neck

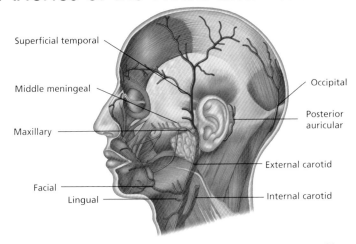

Superficial temporal

Middle meningeal

Maxillary

Facial

Lingual

Occipital

Posterior auricular

External carotid

Internal carotid

The head and neck are supplied with blood from the two common carotid arteries that ascend either side of the neck. These are encased in a protective covering of connective tissue called the carotid sheath. The carotid arteries divide to form the internal and external carotid arteries; the former enters the skull to supply the brain and the latter provides branches to the face and scalp. There are a number of arteries to the face and neck, including the facial artery, which supplies blood to the face and palate and to the lips. Many of the branches of the external carotid artery have a wavy or looped course. This flexibility ensures that when structures (such as the mouth and larynx) move, the vessels are not stretched.

Body system:	cardiovascular system
Location:	head and neck
Function:	provide blood rich in oxygen and nutrients to the tissues of the face and neck
Components:	carotid, superficial temporal, occipital, maxillary, meningeal, facial, lingual, posterior auricular arteries
Related parts:	structures in the head, connecting blood vessels, heart

Veins of the Head and Neck

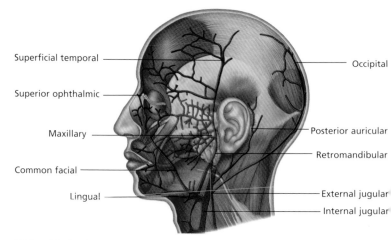

Superficial temporal

Superior ophthalmic

Maxillary

Common facial

Lingual

Occipital

Posterior auricular

Retromandibular

External jugular

Internal jugular

Blood drains from the head and neck back to the heart via three main pairs of veins: the vertebral veins, which run through the vertebrae of the neck, and the internal and external jugular veins. These large veins divide in the neck to form numerous smaller ones that branch out over the head. The veins have a similar distribution as the arteries, and many of them also share the same names. Unlike veins in the rest of the body, those in the head and neck do not have one-way valves, and blood therefore returns to the heart by gravity. Two of the major branches in the head are the facial and retromandibular veins. Between them, these two veins drain venous blood from most of the face and scalp.

Body system :	cardiovascular system
Location:	head and neck
Function:	drain deoxygenated blood from the tissues of the face and neck and return it to the heart
Components:	jugular, vertebral, superficial temporal, occipital, maxillary, superior ophthalmic, facial, lingual, retromandibular, posterior auricular
Related parts:	structures in the head, connecting blood vessels, heart

Eyeball

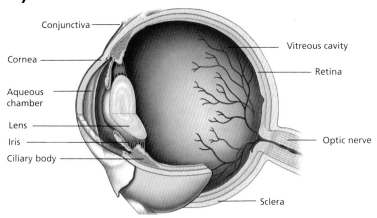

Conjunctiva

Cornea

Aqueous
chamber

Lens

Iris

Ciliary body

Vitreous cavity

Retina

Optic nerve

Sclera

The eyes are specialized organs of sight designed to detect patterns of light. Each eyeball sits is a bony cavity in the skull, embedded in protective fatty tissue and is divided into three internal chambers. The anterior and posterior aqueous chambers are at the front of the eye and are separated by the iris (the colored part of the eye). These chambers are filled with clear, watery substance called aqueous humor, which is secreted by the ciliary body. The largest of the chambers is the vitreous cavity, which lies behind the aqueous chambers and is separated from them by the lens and the suspensory ligaments. The vitreous cavity is filled with clear, jellylike vitreous humor.

Body system:	special senses
Location:	in the head, either side of the nose
Function:	channel light through a series of refractive media to light-sensitive cells in the retina
Components:	cornea, sclera, conjunctiva, iris, lens, anterior and posterior aqueous chambers, vitreous cavity, retina, optic nerve
Related parts:	optic nerve, visual cortex in brain

Layers of the Eye

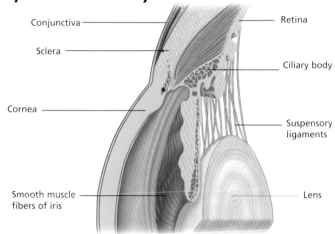

Conjunctiva

Sclera

Cornea

Smooth muscle fibers of iris

Retina

Ciliary body

Suspensory ligaments

Lens

The eyeball is covered by three outer layers, each of which has a different function. The outermost protective layer, the sclera, is tough and fibrous and is visible as the "white of the eye." As this outer layer passes over the iris, it becomes transparent to allow light to enter the eye and is known as the cornea. The sclera and cornea at the front of the eye are covered by a protective membrane called the conjunctiva. The middle layer, the uvea, contains many blood vessels, nerves, and pigmented cells and is divided into three main regions: the choroid, the ciliary body, and the iris. The innermost layer of the eye, the retina, is a layer of nerve tissue containing light-sensitive cells that line the vitreous chamber.

Body system:	special senses
Location:	outermost layers of the eyeball
Function:	protect eye, allow light to pass through to reach the retina (light-sensitive cells at the back of the eye)
Components:	sclera, cornea, conjunctiva, uvea, retina
Related parts:	eyelids, optic nerve, visual cortex in brain

Muscles of the eye

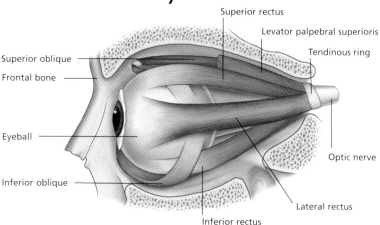

Superior rectus

Levator palpebral superioris

Tendinous ring

Superior oblique

Frontal bone

Eyeball

Inferior oblique

Optic nerve

Lateral rectus

Inferior rectus

The rotational movements of the eye are controlled by six ropelike extraocular muscles, which attach directly to the sclera. Four of the muscles are rectus (straight) muscles—superior, inferior, lateral (the temple side of the eye), and medial (nasal side)—and arise from a band of tough fibrous tissue at the back of the eye called the tendinous ring. The remaining two extraocular muscles are oblique muscles. A further muscle, the levator palpebrae superioris, is located in the orbit (socket) and moves the upper eyelid. Each of the muscles has a different role; for example, the inferior oblique muscle raises the eye and rotates it outward, whereas the eye is pulled upward and outward by the superior rectus muscle.

Body system:	musculoskeletal system
Location:	attached to the sclera of the eye
Function:	move the eyes upward, downward, and from side to side
Components:	superior and inferior oblique, superior and inferior rectus, lateral and medial rectus, levator palpebrae superioris, tendinous ring
Related parts:	eyeball, tendinous ring

Movement of the Eye

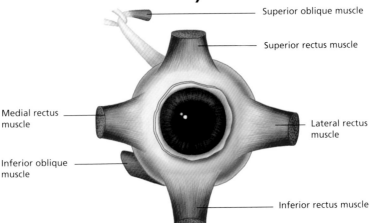

Superior oblique muscle

Superior rectus muscle

Medial rectus muscle

Lateral rectus muscle

Inferior oblique muscle

Inferior rectus muscle

The contraction of the extraocular muscles surrounding the eye is controlled by cranial nerves, specifically the trochlear (CNIV), oculomotor (CNIII) and abducens (CNVI). The muscles act individually to move the eyeball, however, the direction of turn for a particular muscle differs between right and left eyes. For example, in the right eye, the lateral rectus muscle will turn the eye to the right, while in the left eye it would turn the eye to the left. Eye movements normally occur in parallel, so different muscles in each eye act together to turn the eyes. For example, to look left, the lateral rectus muscle will turn the left eye and the medial rectus muscle the right eye.

Body system :	special senses
Location:	attached to the sclera of the eye
Function:	move the eyes upward, downward, and from side to side
Components:	superior and inferior oblique, superior and inferior rectus, lateral and medial rectus
Related parts:	eyeball, tendinous ring

Eyelids

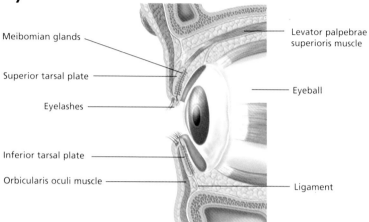

Meibomian glands

Superior tarsal plate

Eyelashes

Inferior tarsal plate

Orbicularis oculi muscle

Levator palpebrae
superioris muscle

Eyeball

Ligament

The eyelids are folds of skin that close over the eye to protect it from bright light or injury. Each lid is strengthened by a band of dense elastic connective tissue called a tarsal plate, which gives the eyelids a curvature to match the eye. The inner and outer ends of each plate are attached to the underlying bone by tiny ligaments. The tarsal plates contain small meibomian glands. These secrete an oily liquid that prevents the lids from sticking together. Eyelashes project from the free edges of the lids and these help to stop foreign bodies from entering the eye. The eye closes due to movement of the upper lid—the orbicularis orbis muscle contracts to close the eye and the lid is opened by the levator palpebrae superioris muscle.

Body system:	special senses
Location:	above and below the eye
Function:	protect the eye from bright light and injury; the lids also spread lubricating tears across the eye
Components:	folds of skin, tarsal plates
Related parts:	orbicularis oculi and levator palpebrae superioris muscles

Conjunctiva

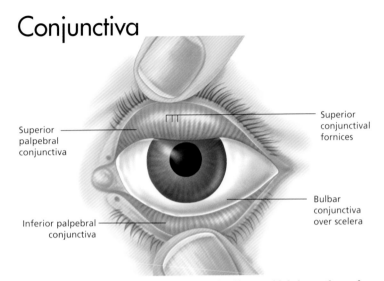

Superior palpebral conjunctiva

Superior conjunctival fornices

Bulbar conjunctiva over sclera

Inferior palpebral conjunctiva

The conjunctiva is a very thin membrane that lines and lubricates the surface of the eyeball and the inner surfaces of the eyelid. It has two parts: the bulbar conjunctiva covers the sclera (the white of the eye) and the palpebral conjunctiva lines the inner surface of both eyelids. The bulbar conjunctiva is thin and transparent and is separated from the sclera by loose connective tissue. It does not cover the cornea (which lies over the iris and pupil) but attaches to its edge. The palpebral conjunctiva forms deep recesses, known as conjunctival fornices, where it meets the bulbar conjunctiva. Normally, the membrane is colorless, but its tiny blood vessels can become inflamed and pink due to irritation or infection.

Body system:	special senses
Location:	covering the eyeball and the inner surfaces of the eyelids
Function:	protects and lubricates the vulnerable surface of the eye
Components:	bulbar conjunctiva, palpebral conjunctiva
Related parts:	other parts of the eye

Lacrimal Apparatus

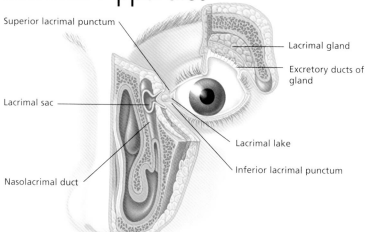

Superior lacrimal punctum

Lacrimal gland

Excretory ducts of gland

Lacrimal sac

Lacrimal lake

Inferior lacrimal punctum

Nasolacrimal duct

The eyes are kept moist by the continuous production of lacrimal fluid (tears) by the lacrimal glands. This fluid also contains an antibacterial substance that protects against infection. Most of this fluid is lost through evaporation, and the remainder is drained to the back of the nose. The lacrimal glands lie just above the outer side of the eye within a recess in the bony socket. They each have up to 12 ducts, which carry fluid away from the gland and to the conjunctival sac through openings under the upper lid. After traveling across the surface of the eye, the tears collect in the lacrimal lake at the inner corner. Excess fluid passes through tiny openings called puncti, and down to the nose through the nasolacrimal ducts.

Body system:	special senses
Location:	above and within the eye, and in the nasal cavity
Function:	keeps the eye moist and provides protection against bacterial infection
Components:	lacrimal glands, lacrimal lakes, puncti, nasolacrimal ducts
Related parts:	eyes, nose

Nose

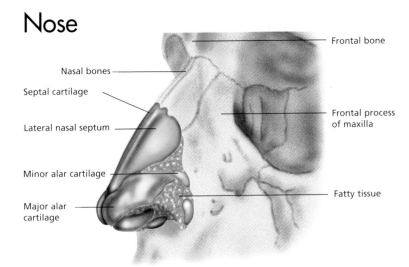

Frontal bone

Nasal bones

Septal cartilage

Lateral nasal septum

Frontal process of maxilla

Minor alar cartilage

Major alar cartilage

Fatty tissue

The external nose is a pyramid-shaped structure in the center of the face. The underlying nasal cavity is a relatively large space and is the very first part of the respiratory tract. The upper part of the nose is made up of bone and the lower section is formed from cartilage and fibrous tissue. The bridge of the nose consists almost entirely of the two nasal bones, which join at their upper edge to the frontal bone of the skull and at their sides to the maxillae (upper jaw bones). The lower part of the nose is made up of plates of cartilage, one of which forms the shape of the nostrils (the external openings of the nose). The nostrils are separated by the septal cartilage and are lined with mucous membrane and tiny hairs that filter inhaled air.

Body system :	respiratory system
Location:	center of the face; nasal cavity extends backward into the skull
Function:	filters, warms, and moistens inhaled air; detects smells
Components:	nasal bones, septal cartilage, nasal cartilage, alar cartilage, nostrils
Related parts:	facial bones, skull

Nasal Cavity

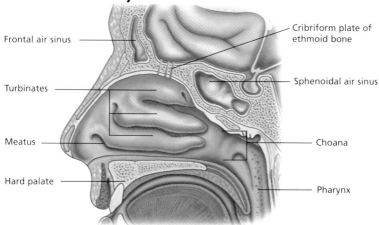

Frontal air sinus

Turbinates

Meatus

Hard palate

Cribriform plate of ethmoid bone

Sphenoidal air sinus

Choana

Pharynx

The mucous-lined nasal cavity runs from the nostrils to the pharynx and is divided in two by a vertical plate called the nasal septum, which is part bone and part cartilage. Each half of the nasal cavity is open at the front at the nostril and at the back into the pharynx through an opening called the choana. The roof of the nasal cavity is arched from front to back—the central part of this roof is the cribriform plate of the ethmoid bone, a strip of bone perforated with a number of holes. This bone forms part of the cranial cavity, which houses the brain. The side walls of the nasal cavity have three horizontal projections called turbinates and below each turbinate is a space called a meatus.

Body system:	respiratory system
Location:	extends from the nostrils to the pharynx (throat)
Function:	moistens, filters, and warms inhaled air; detects smell
Components:	nostrils, cribriform plate, upper, middle, and lower meatus, upper, middle, and lower turbinates, choana
Related parts:	pharynx, skull, nasolacrimal ducts

Paranasal Sinuses

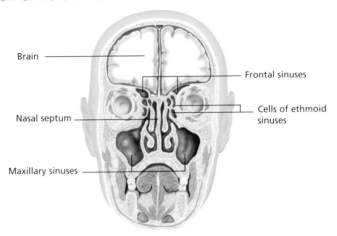

Brain

Frontal sinuses

Cells of ethmoid sinuses

Nasal septum

Maxillary sinuses

The paranasal sinuses form a complex system of air-filled cavities in the bones around the nasal cavity. The four pairs of paranasal sinuses—the maxillary, ethmoidal, frontal, and sphenoidal (not shown) sinuses—are named according to the bones in which they are situated. Each sinus is lined by cells that secrete mucus and opens into the nasal cavity through a tiny opening called an ostium, through which the mucus drains. It is thought that one function of the sinuses is to add to the resonance of the voice. The sinuses are also believed to act as thermal insulators by preventing cold inhaled air from cooling the brain tissue. A third function of the sinuses is believed to be to lighten the weight of the skull.

Body system:	respiratory system
Location:	in the bones surrounding the nasal cavity
Function:	add resonance to the voice, lighten the weight of the skull, prevent cold air from cooling the brain, produce mucus
Components:	maxillary, ethmoidal, frontal, and sphenoidal sinuses
Related parts:	nasal cavity, skull, brain

Inside the Sinuses

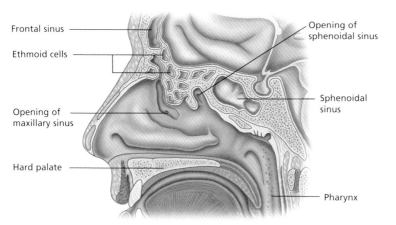

Frontal sinus

Ethmoid cells

Opening of maxillary sinus

Hard palate

Opening of sphenoidal sinus

Sphenoidal sinus

Pharynx

The sphenoidal sinuses are located behind the roof of the nasal cavity, within the sphenoidal bone. The two sinuses lie side by side, separated by a thin, bony partition, and open into the uppermost part of the nasal cavity. The ethmoidal sinuses are situated between the thin, inner wall of the eye socket and the side wall of the nasal cavity. Unlike other paranasal sinuses, these are made up of multiple communication cavities called ethmoid air cells, which drain into the upper and middle nasal cavity. The maxillary sinuses are the largest sinuses and are situated within the maxillae (upper jaw bones). Infections and inflammation are more common here than in the other sinuses due to inefficient drainage of secretions.

Body system:	respiratory system
Location:	in the bones surrounding the nasal cavity
Function:	add resonance to the voice, lighten the weight of the skull, prevent cold air from cooling the brain, produce mucus
Components:	maxillary, ethmoidal, frontal, and sphenoidal sinuses
Related parts:	nasal cavity, skull, brain

Oral Cavity

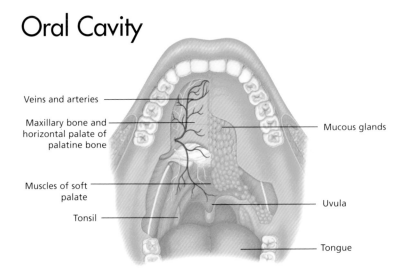

Veins and arteries

Maxillary bone and horizontal palate of palatine bone

Mucous glands

Muscles of soft palate

Uvula

Tonsil

Tongue

Also known as the mouth, the oral cavity extends from the lips to the fauces (the opening leading to the pharynx). The roof of the mouth, shown here in layers, shows two distinct structures: the dental arch and the palate. The dental arch is the curved part of the maxilla (upper jaw bone) at the front and sides of the roof, and the palate is a soft plate of tissue that separates the mouth from the nose. The front two thirds of the palate is bony and hard, while the rear third is soft, consisting of mucous glands and muscle. These muscles close off the nasal cavity from the mouth during swallowing. At the back of the oral cavity are located the tonsils (two rings of lymphoid tissue) and the uvula, a pendulous extension of the soft palate.

Body system:	digestive system
Location:	extends from the lips to the throat
Function:	provides passageway for air and food, houses tongue and teeth, produces saliva to begin process of digestion
Components:	lips, teeth, tongue, soft palate, hard palate, tonsils, uvula
Related parts:	salivary glands, pharynx, nasal cavity, airways

Floor of the Mouth

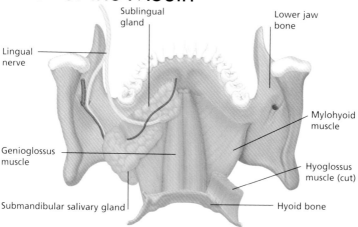

Sublingual gland

Lower jaw bone

Lingual nerve

Mylohyoid muscle

Genioglossus muscle

Hyoglossus muscle (cut)

Submandibular salivary gland

Hyoid bone

The boundaries of the floor of the mouth are formed by the lower jaw and the teeth. The floor itself acts as a foundation for a network of muscles and glands that are essential to the function of the mouth. The tongue (not shown here) is attached to, and lies over, the mylohyoid muscle, which forms the muscular floor of the mouth. The hyoglossus muscle anchors the tongue to the hyoid bone (involved in swallowing) and provides extra strength, while the genioglossus muscle prevents the tongue from falling back and blocking the airway. There are a pair of submandibular and sublingual salivary glands on either side of the oral floor, and these provide a regular flow of saliva into the mouth.

Body system:	digestive system
Location:	extends from the lips to the throat
Function:	provides passageway for air and food, houses tongue and teeth, produces saliva, which begins process of digestion
Components:	soft palate, hard palate, salivary glands, tonsils, uvula
Related parts:	lips, teeth, tongue, airways

Tongue

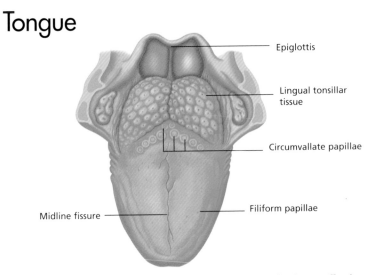

Epiglottis

Lingual tonsillar tissue

Circumvallate papillae

Filiform papillae

Midline fissure

The numerous functions of the tongue include speech, mastication, swallowing and providing the sense of taste. The tongue is basically a mass of muscle whose upper surface is covered by numerous filiform papillae, tiny protuberances that give the surface a rough feel. Scattered among them are larger fungiform and circumvallate papillae—it is in the circumvallate papillae that the taste buds are largely situated. The posterior third of the tongue's surface has a cobbled appearance due to the presence of 40–100 nodules of lymphoid tissue, which together form the lingual tonsil, a ring of tissue that protects the throat. At the very back of the tongue is the epiglottis, which seals off the airway during swallowing.

Body system:	digestive system
Location:	the oral cavity
Function:	speech, mastication (chewing), swallowing, taste
Components:	muscle, papillae, lymphoid tissue
Related parts:	teeth, oral cavity, pharynx (throat)

Taste Buds

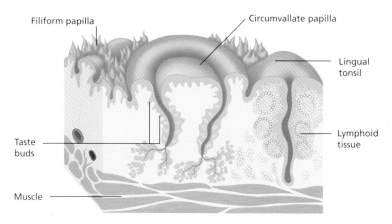

Filiform papilla

Circumvallate papilla

Lingual tonsil

Taste buds

Lymphoid tissue

Muscle

The taste buds are nests of cells sensitive to flavored substances in solution. These cells are located mainly in the furrows of structures called circumvallate papillae toward the rear of the tongue. There are, however, taste buds scattered over the rest of the tongue's surface and in the lining of the cheeks and throat. It is traditional to describe tastes as either salty, sweet, bitter, or sour, but the central processing of taste data by the brain is complex. It seems that when a nerve fiber is carrying data from a taste bud, it is responding to several or all four of these basic taste sensations with differing sensitivities. Furthermore, the sense of taste is interrelated with the sense of smell, so food becomes tasteless with a heavy cold.

Body system:	digestive system
Location:	mainly in the furrows of circumvallate papillae
Function:	detect taste
Components:	clusters of flavor-sensitive cells
Related parts:	other areas of tongue, brain, nose

Muscles of the Tongue

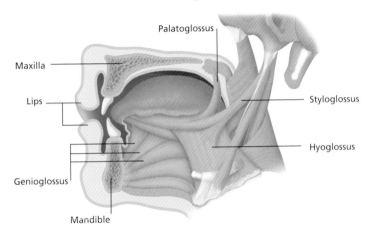

Palatoglossus

Maxilla

Lips

Styloglossus

Hyoglossus

Genioglossus

Mandible

There are two types of muscle involved in movement of the tongue: intrinsic and extrinsic muscles. Intrinsic muscles are those inside the tongue and consist of groups of fiber bundles running the length, breadth, and depth of the organ. The intrinsic muscles are mainly responsible for altering the shape of the tongue. The extrinsic muscles—the genioglossus, hyoglossus, styloglossus, and the palatoglossus—enter the tongue from origins outside it and have the ability to change its position. This variety of muscles makes the tongue an extremely mobile organ that is able to assist with mastication (chewing) and swallowing, as well as change its size and position to act as a resonator during speech.

Body system:	digestive system
Location:	within and surrounding the tongue
Function:	assist in speech, chewing, and swallowing
Components:	intrinsic tongue muscles, genioglossus, hyoglossus, styloglossus, palatoglossus
Related parts:	oral cavity, teeth, pharynx (throat)

Lymph Drainage of the Tongue

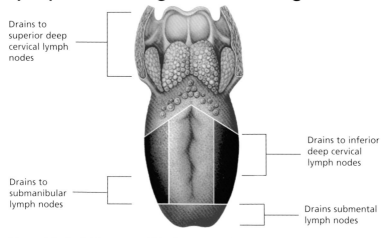

Drains to superior deep cervical lymph nodes

Drains to inferior deep cervical lymph nodes

Drains to submanibular lymph nodes

Drains submental lymph nodes

Lymph is the fluid present within the vessels of the lymphatic system. Its main function is to collect excess fluid from the tissues and return it to the blood circulation. The lymph vessels of the tongue have their own drainage pattern. Lymph from both sides of the tip of the tongue drains into the submental group of nodes, which are located under the chin. The submandibular nodes lie under the jaw and receive lymph from the sides of the tongue. The central area of the tongue drains to the inferior (lower) deep cervical nodes, which lie alongside the internal jugular veins deep within the neck. The superior (upper) cervical lymph nodes receive lymph from the the back of the tongue.

Body system:	lymphatic system
Location:	lymph nodes in the neck
Function:	drain excess fluid from the tissues of the tongue
Components:	lymphatic vessels and submental, submandibular, and cervical nodes
Related parts:	other parts of the lymphatic system

Teeth

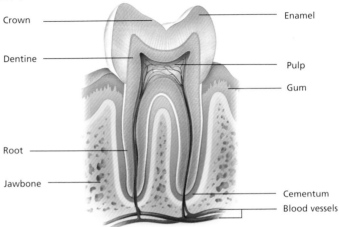

Crown — Enamel

Dentine — Pulp

Gum

Root

Jawbone

Cementum

Blood vessels

The teeth are hard, conical structures partly embedded in the jawbones, which are designed to bite off and chew solid food. The visible part of the tooth is known as the crown and is composed of calcified material called dentine, which is in turn covered with enamel, the hardest substance in the body. The root of each tooth is embedded in a socket (alveolus) in the jaw and is covered by a layer of cementum which, together with the periodontal ligaments, anchors the root to the bone. Inside each tooth is an internal pulp cavity containing soft connective tissue, blood vessels, and nerves. In adults, each side (or quadrant) has eight teeth—two incisors, one canine, two premolars, and three molars—making 32 in total.

Body system:	digestive system
Location:	oral cavity
Function:	biting and chewing
Components:	enamel, dentine, connective tissue, blood vessels, nerves
Related parts:	oral cavity, jawbone

Development of the Teeth

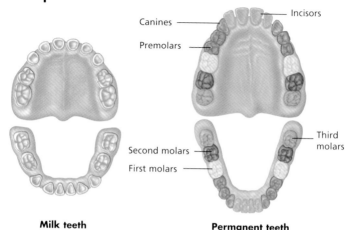

Incisors

Canines

Premolars

Third molars

Second molars

First molars

Milk teeth

Permanent teeth

There are two major phases of tooth development during childhood. This is to allow the head to grow and adult teeth to develop. Around the sixth week of pregnancy, teeth begin to develop in the human embryo. Six to eight weeks after birth, root growth pushes the tooth crown through the gum in the process called teething. This first set are the primary or deciduous teeth (milk teeth). They erupt in a specific order, usually the lower central incisors first, then the upper central incisors. Tooth buds for the permanent teeth develop at the same time as deciduous teeth are growing, but they remain dormant until age five to seven, when they begin to grow. Gradually, the deciduous teeth are shed as the permanent teeth appear.

Body system:	digestive system
Location:	oral cavity
Function:	biting and chewing
Components:	enamel, dentine, connective tissue, blood vessels, nerves
Related parts:	oral cavity, jaw bone

Salivary Glands

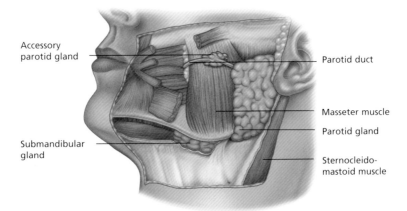

Accessory parotid gland

Parotid duct

Masseter muscle

Parotid gland

Submandibular gland

Sternocleido-mastoid muscle

Between them the salivary glands produce about 1½ pints (¾ L) of saliva a day. Saliva plays a major role in lubricating and protecting the mouth and teeth, as well as helping to chew and swallow food. There are three pairs of major glands—the parotid (the largest gland located just in front of the ear), the submandibular, and the sublingual glands—which provide about 90 percent of the saliva, while the remainder is produced by minor glands in the cheeks, lips, and palate. There are two distinct types of saliva-producing cells located at the end of a branching series of ducts in the glands. Mucous cells form a viscous mucin-rich product and serous cells produce a watery fluid containing the enzyme amylase.

Body system :	digestive system
Location:	surrounding the oral cavity
Function:	produce saliva, which is important for lubrication and assisting with the initial breakdown of food
Components:	parotid, submandibular, sublingual glands
Related parts:	oral cavity, some arteries and nerves

Submandibular and Sublingual Glands

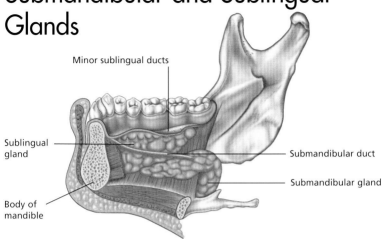

Minor sublingual ducts

Sublingual gland

Body of mandible

Submandibular duct

Submandibular gland

The two smaller pairs of salivary glands are the submandibular and the sublingual glands, which are situated in the floor of the mouth. The submandibular glands are about the size of a walnut and have two parts, a large superficial part and a smaller deep part. The saliva produced in these glands drains to the undersurface of the tongue via the submandibular ducts. The sublingual glands are the smallest of the major salivary glands and lie underneath the tongue in the sublingual fossa, almost meeting in the midline. Unlike the other glands, the sublingual glands do not have a major collecting duct and instead have many smaller ones that open into the floor of the mouth or into the submandibular duct.

Body system:	digestive system
Location:	surrounding the oral cavity
Function:	produce saliva which is important for lubrication and assisting with the initial breakdown of food
Components:	parotid, submandibular, sublingual glands
Related parts:	oral cavity, some arteries and nerves

Infratemporal Fossa

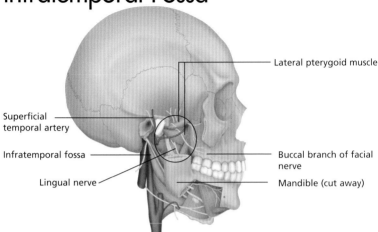

Lateral pterygoid muscle

Superficial temporal artery

Infratemporal fossa

Lingual nerve

Buccal branch of facial nerve

Mandible (cut away)

The infratemporal fossa (fossa means a depression or hollow) is a cavity at the side of the skull that houses a number of important nerves, blood vessels and muscles involved in mastication (chewing). The fossa is located in the skull beneath the cheekbone and in front of the ear. The structures within the infratemporal fossa include the pterygoid muscles, which assist in opening the jaw, the pterygoid venous plexus (a network of vessels that surrounds the muscles) and the maxillary artery. In addition, several important nerves that supply the face run through the fossa, namely branches of the facial and mandibular nerves and the otic ganglion (part of the autonomic nervous system).

Body system:	musculoskeletal system
Location:	depression at the side of the head
Function:	houses nerves, blood vessels, and muscles
Components:	area surrounded by bones of the skull
Related parts:	pterygoid muscles, mandibular and facial nerves, maxillary artery, pterygoid venous plexus

Pterygopalatine Fossa

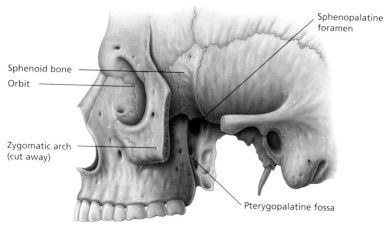

Sphenopalatine foramen

Sphenoid bone

Orbit

Zygomatic arch (cut away)

Pterygopalatine fossa

Just behind the maxilla (upper jaw bone) is a small funnel-shaped space called the pterygopalatine fossa. The fossa is a distribution center that communicates with (opens into) all of the main regions of the head, such as the mouth, nose, eyes, face, infratemporal fossa, and also with the brain. The pterygopalatine fossa contains important nerves and blood vessels, including the maxillary artery and nerve and the pterygopalatine ganglion, all of which enter and exit the region through the sphenopalatine foramen. The pterygopalatine ganglion is a point of convergence for different nerves, which unite within the pterygopalatine fossa, and it is a relay station for nerve fibers that control glandular secretions.

Body system:	musculoskeletal system
Location:	behind the maxilla (upper jaw bone)
Function:	houses nerves and blood vessels
Components:	area surrounded by bones of the skull
Related parts:	maxillary artery and nerve, pterygopalatine ganglion

Maxillary Artery

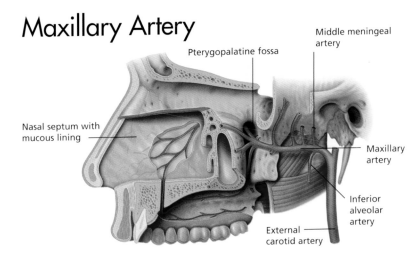

The maxillary artery is the larger of the two terminal branches of the external carotid artery in the neck. Almost immediately, numerous branches, such as the inferior alveolar artery, the middle meningeal artery, and small branches to the inner ear, leave the main artery, which then passes across to the mandible (jawbone) and runs along the lateral pterygoid muscle to the pterygopalatine fossa. Once in the fossa, the artery divides into many smaller vessels, which supply oxygenated blood to the upper teeth and lip, the nasal cavity, the hard and soft palates, the skin of the lower eyelid, the paranasal sinuses, the orbit (eye socket), and the nose. The muscles of mastication (chewing) are also supplied by the maxillary artery.

Body system:	cardiovascular system
Location:	terminal branch of external carotid artery
Function:	supplies oxygen-rich blood to many of the structures in the head
Components:	many deep and superficial branches
Related parts:	face, nasal cavity, hard and soft palates, maxillary teeth, sinuses, inner ear, external carotid artery, heart

Ear

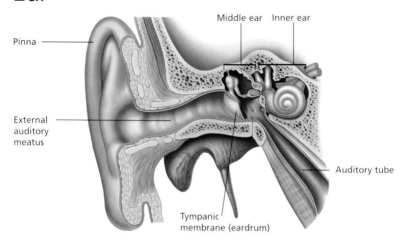

Middle ear · Inner ear

Pinna

External auditory meatus

Auditory tube

Tympanic membrane (eardrum)

The ear can be anatomically divided into three parts: the external and middle ear, which gather and transmit sound, and the inner ear, which is the organ of hearing and balance. The external ear consists of the visible part (the pinna), which is made up of skin and cartilage, and the external auditory meatus, which channels sound waves toward the middle ear. Inside the external meatus are tiny hairs and ceruminous glands that secrete cerumen (wax). This combination of wax and hairs helps prevent dust and foreign bodies from entering the ear. At the inner end of the meatus is the tympanic membrane, or eardrum, which vibrates in response to sound waves and marks the border between the external and middle ear.

Body system:	special senses
Location:	on either side of the head, extending inward
Function:	gather and transmit sound waves; organ of hearing and balance
Components:	external, middle, and inner ear
Related parts:	skull, vestibulocochlear nerve, brain

Middle Ear

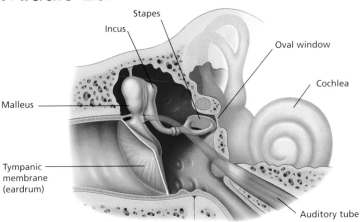

Stapes

Incus

Oval window

Cochlea

Malleus

Tympanic membrane (eardrum)

Auditory tube

Beyond the tympanic membrane (eardrum) is the middle ear, an air-filled cavity that helps transmit sound to the inner ear and is connected to the pharynx (throat) via the auditory (eustachian) tube. Within the middle ear are three tiny bones called ossicles that are linked together in such a way that movements of the eardrum are transmitted across them to the oval window at the entrance to the inner ear. The first bone, the malleus, is attached to the inner surface of the eardrum at one end, and to the second bone, the incus, at the other. The stapes is the third ossicle and is connected both to the incus and to the oval window. All three ossicles are held in place by tiny ligaments.

Body system:	special senses
Location:	between the eardrum and the inner ear
Function:	transmits sound in the form of vibrations from the eardrum to the inner ear; maintains appropriate pressure within the ear
Components:	malleus, incus, stapes, auditory tube
Related parts:	external ear, inner ear, pharynx

Inner Ear

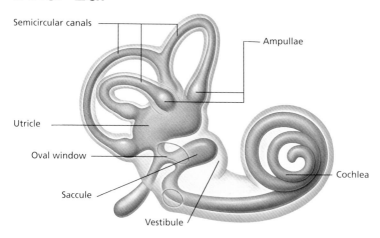

Semicircular canals

Ampullae

Utricle

Oval window

Saccule

Vestibule

Cochlea

This part of the ear contains the organs of balance and hearing. The inner ear has two divisions: the outer bony labyrinth is a system of channels and has three regions called the vestibule, the semicircular canals, and the cochlea; the inner membranous labyrinth consists of a series of linked sacs or ducts contained within the bony labyrinth. The vestibule is the central part of the bony labyrinth and contains two membranous sacs, the saccule and the utricle, which provide information about head position, while the semicircular canals contain receptors that detect head movements. The cochlea is a bony, spiral canal that houses the organ of hearing, known as the organ of Corti.

Body system:	special senses
Location:	deep in the skull behind the orbit (eye socket)
Function:	contains organs of hearing and balance
Components:	outer bony labyrinth containing vestibule, semicircular canals, and cochlea, inner membranous labyrinth
Related parts:	external and middle ear, vestibulocochlear nerve, brain

Inside the Neck

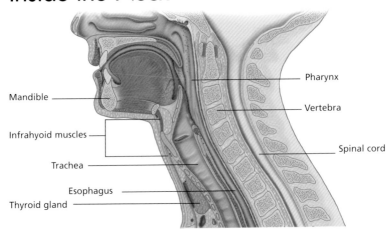

Pharynx

Mandible

Vertebra

Infrahyoid muscles

Spinal cord

Trachea

Esophagus

Thyroid gland

The neck is defined as the region lying between the bottom of the lower jaw and the top of the clavicle (collarbone). Within this relatively small area there are numerous vital structures that are closely packed together between layers of connective tissue, which serve to protect and anchor them. Both the trachea (main airway) and the esophagus (gullet) travel downward through the neck from the mouth to join the lungs and stomach respectively. Wrapped around the trachea is the thyroid gland, which secretes hormones that help control the body's internal environment. The most important structure is the spinal cord, which is encased in, and protected by, the cervical vertebrae.

Body system	various
Location:	between the lower jaw and the clavicle (collarbone)
Function:	connects the head to the body, allowing movement of the head; houses important structures, such as the trachea and esophagus
Components:	skin, muscle, connective tissue, spinal cord, vertebrae, trachea, esophagus, thyroid gland
Related parts:	head, thorax

Muscles of the Neck

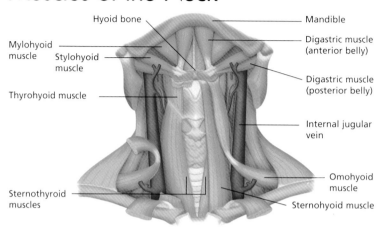

Hyoid bone

Mandible

Mylohyoid muscle

Digastric muscle (anterior belly)

Stylohyoid muscle

Thyrohyoid muscle

Digastric muscle (posterior belly)

Internal jugular vein

Omohyoid muscle

Sternothyroid muscles

Sternohyoid muscle

Two groups of muscles—the suprahyoid and infrahyoid muscles—run down the front of the neck from the mandible (jaw) to the sternum (breastbone). These muscles are responsible for the movements of the jaw, the hyoid bone, and the larynx (voicebox) and are of particular importance in swallowing. The paired suprahyoid muscles lie between the jaw and the hyoid bone. They include the digastric muscle, which has two "bellies" connected by a tendon, the stylohyoid, the geniohyoid, and the mylohyoid, which forms the floor of the mouth. The strap-shaped infrahyoid muscles are located between the hyoid bone and the sternum and include the sternohyoid, omohyoid, thyrohyoid, and sternothyroid muscles.

Body system:	musculoskeletal system
Location:	run longitudinally down the front of the neck
Function:	produce movement in the jaw, hyoid bone, and larynx (voicebox) and are particularly important in swallowing
Components:	suprahyoid and infrahyoid muscle groups
Related parts:	other structures in the neck, mandible, clavicle, sternum

Scalene and Prevertebral Muscles

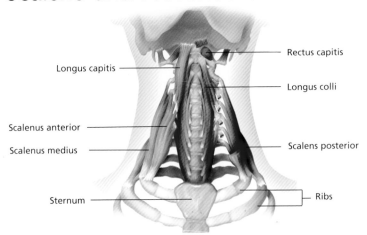

Longus capitis

Rectus capitis

Longus colli

Scalenus anterior

Scalenus medius

Scalens posterior

Sternum

Ribs

The center of gravity of the head lies in front of the spine, and the muscles and ligaments of the back and neck must work hard to prevent the head from falling forward. Much of the forward and lateral flexion of the head is achieved by the coordinated action of the neck flexor muscles: the scalenes, the prevertebrals, and the powerful sternocleidomastoids. The three scalene muscles run from either side of the cervical vertebrae to attach to the first and second ribs. These muscles also help to raise the first two ribs during inspiration (breathing in). The prevertebral muscles lie in front of the cervical vertebrae, some extending from the skull to the first cervical vertebrae and others down to the junction of the neck and chest.

Body system:	musculoskeletal system
Location:	scalenes run from the cervical vertebrae to the ribs; prevertebrals run from the skull to the cervical vertebrae or sternum
Function:	keep head stable and erect on the spine, enable flexion of the head and neck and raise the first two ribs during inspiration
Components:	scalenus anterior, scalenus medius, scalenus posterior, longus capitis, and longus colli muscles
Related parts:	skull, vertebrae, ribs

Sternocleidomastoid Muscles

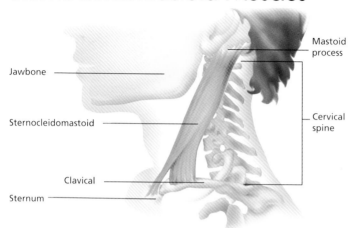

Mastoid process

Jawbone

Cervical spine

Sternocleidomastoid

Clavical

Sternum

The sternocleidomastoid muscles are the main head flexor muscles. These powerful muscles can be seen very prominently under the skin on either side of the front of the neck. They run from the mastoid process (a prominence on the base of the skull) downward to the sternum (breastbone) and clavicle (collarbone). At this lower end, each muscle splits into two segments; one part attaches to the front of the upper sternum, while the second deeper part attaches to the clavicle. The sternocleidomastoid muscles act with the other neck flexor muscles to flex and rotate the neck. Acting alone, each muscle can turn the head toward the opposite shoulder, or tilt or laterally flex the head.

Body system:	musculoskeletal system
Location:	run from mastoid process to sternum and clavicle
Function:	tilt and flex head, flex and rotate neck
Components:	two-headed skeletal muscle
Related parts:	skull, sternum, clavicle, other head flexor muscles

Pharynx

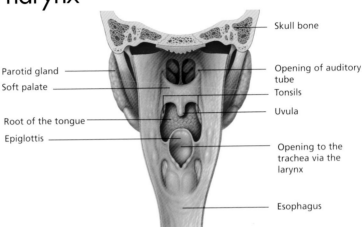

Skull bone

Parotid gland

Soft palate

Opening of auditory tube

Tonsils

Uvula

Root of the tongue

Epiglottis

Opening to the trachea via the larynx

Esophagus

The pharynx is a muscular tube at the back of the throat that acts as a passage for food to the esophagus (gullet) and air to the trachea (main airway). This view of the opened pharynx seen from behind shows its three parts. The nasopharynx lies above the soft palate. On each side of the nasopharynx is a tubal elevation, the end of the auditory tube that enables air pressure to be equalized between the pharynx and the middle ear. The oropharynx is located at the back of the throat. Its roof is the undersurface of the soft palate and its floor is formed by the back of the tongue. The laryngopharynx extends from the epiglottis to the lower part of the cricoid cartilage, where it becomes the esophagus.

Body system :	respiratory and digestive systems
Location:	runs from the nasal cavity to the top of the esophagus
Function:	acts as a passage for both air and food
Components:	nasopharynx, oropharynx, laryngopharynx
Related parts:	mouth, nasal cavity, esophagus, trachea

Muscles of the Pharynx

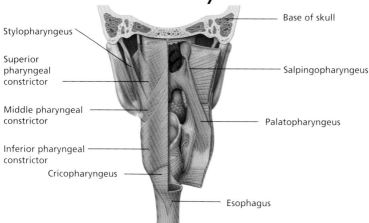

Stylopharyngeus

Superior pharyngeal constrictor

Middle pharyngeal constrictor

Inferior pharyngeal constrictor

Cricopharyngeus

Base of skull

Salpingopharyngeus

Palatopharyngeus

Esophagus

There are six pairs of muscles that make up the pharynx. One group comprises three pairs of constrictor muscles that run across the pharynx: the superior, middle, and inferior constrictors. These constrict (tighten) the pharynx, squeezing food downward into the esophagus. The constrictor muscles overlap each other from below upward (like three stacked cups inside each other) and important structures enter the pharynx in the spaces between them. The other group is formed of three pairs of muscles that run from above the pharynx downward: the salpingopharyngeus, stylopharyngeus, and palatopharyngeus. These raise the pharynx during swallowing, elevating the larynx and protecting the airway.

Body system:	musculoskeletal system
Location:	run across the pharynx, and down into the pharynx from above
Function:	squeeze food downward into the esophagus, raise pharynx during swallowing, elevate larynx, and protect airway
Components:	superior, middle, and inferior constrictors, salpingopharyngeus, stylopharyngeus, and palatopharyngeus muscles
Related parts:	mouth, nasal cavity, esophagus, trachea

Larynx

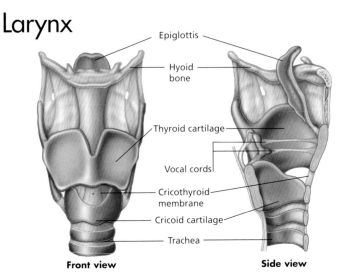

Epiglottis

Hyoid bone

Thyroid cartilage

Vocal cords

Cricothyroid membrane

Cricoid cartilage

Trachea

Front view   **Side view**

The larynx, or voice box, is a short passageway situated in the neck between the pharynx (throat) and the trachea (main airway). The larynx has three important functions: it provides a passageway for air and food; it prevents food and liquid from entering the trachea, using the epiglottis (a spoon-shaped cartilage) to close the passageway during swallowing; it houses the vocal cords and is, therefore, responsible for voice production. The organ is composed of nine cartilages (three single and three paired) connected by membranes, ligaments and muscles. The large single cartilages are the thyroid cartilage (commonly known as the Adam's apple), the epiglottis and the cricoid cartilage.

Body system:	respiratory system
Location:	in the neck between the pharynx and trachea
Function:	provides passageway for food, liquids, and air; prevents food and liquid from entering the airways; responsible for voice production
Components:	nine cartilages (three single and three paired), vocal cords
Related parts:	pharynx, trachea, thyroid gland

Muscles of the Larynx

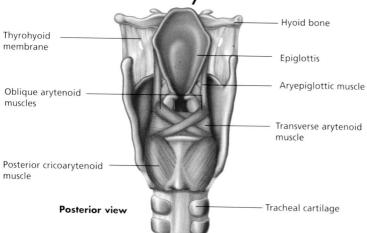

Thyrohyoid membrane

Hyoid bone

Epiglottis

Aryepiglottic muscle

Oblique arytenoid muscles

Transverse arytenoid muscle

Posterior cricoarytenoid muscle

Tracheal cartilage

Posterior view

The muscles of the larynx act to close the entrance to the airways while swallowing to prevent food or liquid from entering the trachea. They also move the vocal cords to enable speech. During swallowing, the epiglottis, along with the rest of the larynx, is raised. As the front surface of the epiglottis hits the rear part of the tongue, it flips backward over the laryngeal inlet. The aryepiglottic muscles are directly attached to the epiglottis and, in conjunction with the oblique arytenoid muscles, help pull it forward. The transverse arytenoid muscle closes the posterior part of the glottis (the gap between the vocal cords), which enables speech formation. The posterior cricoarytenoid muscle opens the glottis.

Body system:	respiratory system
Location:	in the neck between the pharynx and trachea
Function:	close epiglottis to prevent food and fluid from entering the airways; open and close glottis to form speech
Components:	aryepiglottic, transverse arytenoid, oblique arytenoid, posterior cricoarytenoid muscles
Related parts:	vocal cords, epiglottis

Thyroid Gland

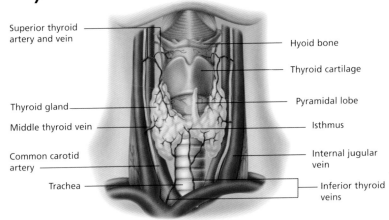

Superior thyroid artery and vein

Hyoid bone

Thyroid cartilage

Pyramidal lobe

Thyroid gland

Isthmus

Middle thyroid vein

Common carotid artery

Internal jugular vein

Trachea

Inferior thyroid veins

The thyroid gland is a large, butterfly-shaped endocrine gland wrapped around the trachea (main airway) just below the larynx (voice box). The gland produces two important hormones—tri-iodothyronine and thyroxine—that play a part in controlling growth and metabolism (chemical processes in the body). The gland also secretes calcitonin, which helps regulate the calcium levels in the blood. The gland, which is composed of hollow, round sacs, has two lobes connected by a central "isthmus." An additional small pyramidal lobe is sometimes present, extending from the isthmus. The superior and inferior thyroid arteries provide the gland with a rich blood supply of about 3–4 oz (80–120 ml) of blood a minute.

Body system:	endocrine system
Location:	wrapped around the trachea (main airway) below the larynx
Function:	secretes hormones involved in metabolism, growth, and calcium regulation
Components:	two lobes and a connecting isthmus
Related parts:	blood circulation, trachea

Parathyroid Glands

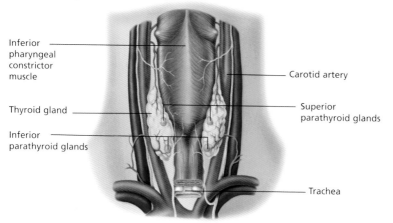

Inferior pharyngeal constrictor muscle

Carotid artery

Thyroid gland

Superior parathyroid glands

Inferior parathyroid glands

Trachea

The tiny, pea-sized parathyroid glands (superior and inferior) are embedded in the posterior surfaces of the lobes of the thyroid gland. There are usually four glands, but the number can vary between individuals. The parathyroid glands are occasionally situated in other parts of the neck or even in the thorax (chest). Each gland contains large numbers of "chief cells" that secrete the hormone parathormone. Together with calcitonin and vitamin D, parathormone controls calcium levels in the body. Parathormone is released when the level of calcium in the blood drops to below normal. It then stimulates bone to release calcium into the bloodstream and enhances the reabsorption of calcium by the kidneys.

Body system:	endocrine system
Location:	usually embedded in the posterior surfaces of the thyroid gland
Function:	produce the hormone parathormone, which regulates calcium levels in the blood
Components:	gland tissue containing secretory "chief cells"
Related parts:	thyroid gland, bones, kidneys

Lymph Drainage of the Head and Neck

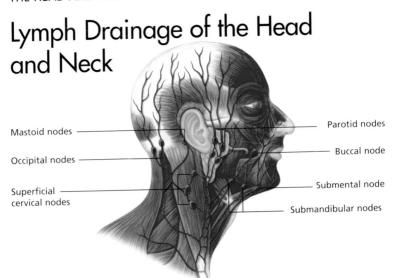

Mastoid nodes

Occipital nodes

Superficial cervical nodes

Parotid nodes

Buccal node

Submental node

Submandibular nodes

The lymphatic system is a network of vessels and nodes throughout the body, whose role is to drain excess fluid from the body's tissues and return it to the blood circulation. Lymph nodes, or glands, filter the fluid as it passes through them, detecting and sometimes destroying foreign bodies. The lymph node groups of the head and neck are named according to nearby structures; for example, the cervical nodes are in the region of the cervical vertebrae and the parotid nodes are adjacent to the parotid salivary gland. The paired lingual tonsils are themselves lymph nodes. Deep within the neck lie other nodes that surround and drain the pharynx (throat), larynx (voice box), and trachea (main airway).

Body system:	lymphatic system
Location:	nodes and vessels scattered around the head and neck
Function:	drain excess fluid away from tissue cells, protects against infection
Components:	numerous vessels, occipital, mastoid, parotid, buccal, sub-mandibular, submental, anterior cervical, superficial cervical nodes
Related parts:	structures in the head and neck

Cervical Vertebrae and Ligaments

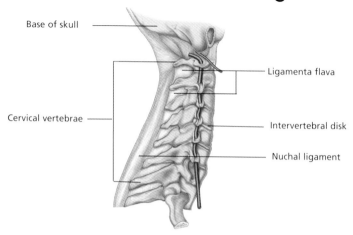

Base of skull

Ligamenta flava

Cervical vertebrae

Intervertebral disk

Nuchal ligament

There are seven cervical vertebrae, which together make up the skeletal structure of the neck. These vertebrae protect the spinal cord, support the skull, and enable neck movement. The first, second, and seventh vertebrae differ structurally from the others. The first, known as the atlas, is a thin ring of bone that articulates directly with the skull. The second—the axis—provides a stable base for the atlas. The seventh vertebra has the largest spine of all the backbones, and it can be easily felt through the skin. The ligaments in the neck have a vital role in binding the cervical vertebrae together while allowing a wide range of movements. They secure the vertebrae to each other and to the skull.

Body system:	musculoskeletal system
Location:	vertebrae located between skull and thoracic vertebrae; ligaments run between vertebrae and up to the skull
Function:	vertebrae protect and support spinal cord, support skull and allow movement; ligaments stabilize and secure the vertebrae
Components:	seven vertebrae, various deep and superficial ligaments
Related parts:	skull, spinal cord, thoracic vertebrae

Vertebral Column

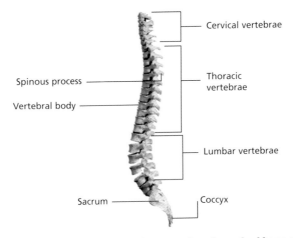

Cervical vertebrae

Thoracic vertebrae

Spinous process

Vertebral body

Lumbar vertebrae

Sacrum

Coccyx

The curved vertebral column, also known as the spine or backbone, extends from the skull to the pelvis. It provides support and stability for the whole body, and is an attachment point for the ribs and the muscles of the back. The vertebral column also houses and protects the delicate spinal cord, which originates at the base of the brain and runs down the central cavity of the column. The 26 bones that make up the column are connected in such a way that they allow flexibility of movement. There are five divisions to the column: the seven cervical vertebrae of the neck; the 12 thoracic vertebrae in the thorax; the five lumbar vertebrae supporting the lower back; and lastly the sacrum and coccyx.

Body system :	musculoskeletal system
Location:	runs from the skull to the pelvis
Function:	provides support and stability and an attachment point for ribs and muscles; allows flexible movement of the back; protects spinal cord
Components:	cervical, thoracic, and lumbar vertebrae, sacrum, coccyx
Related parts:	skull, ribs, pelvis, muscles, and ligaments

Vertebrae

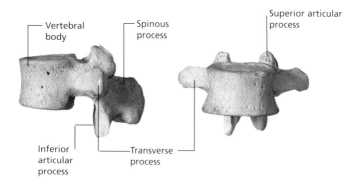

Vertebral body — Spinous process — Superior articular process

Inferior articular process — Transverse process

Although many of the vertebrae in the spine differ in shape and size, all of them have the same basic structure. Each bone consists of a cylindrical body, the weight-bearing area, at the front, a vertebral (neural) arch behind the body and bony projections called processes. Between the body and the vertebral arch is the vertebral foramen (opening) through which the spinal cord passes. Seven processes project outward from the vertebral arch. These include the spinous process and two transverse processes, which provide attachments for muscles and ligaments. Other processes, the superior and inferior articular processes, form joints between the vertebrae to allow movement of the spine.

Body system:	musculoskeletal system
Location:	make up the vertebral column
Function:	provides support and stability and an attachment point for ribs and muscles. Allows flexible movement of the back; protects spinal cord
Components:	body, vertebral arch, and spinous, transverse, superior, and inferior processes
Related parts:	skull, ribs, pelvis, muscles, and ligaments

Intervertebral Disks

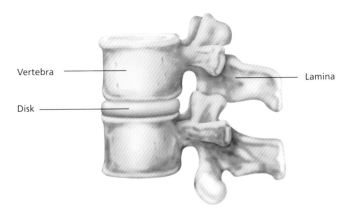

Vertebra

Lamina

Disk

Between each vertebra (backbone) there is a flat disk-shaped structure that cushions and protects the bones and acts as the spine's shock absorber during walking, running, and jumping. Compression of the disks also allows slight movement between the individual bones; collectively this enables the spine to bend considerably over its full length. Each disk is formed of connective tissue with a soft central component (the jellylike nucleus pulposus), which provides the disk with its elasticity and allows its compression. Surrounding the nucleus pulposus is the annulus fibrosis, a tough layer of fibrous tissue, which limits the expansion of the nucleus pulposus and resists any tension in the spine.

Body system:	musculoskeletal system
Location:	between each vertebra in the vertebral column
Function:	act as shock absorbers during movement; allow the spine flexibility to flex and bend
Components:	central nucleus pulposus and outer fibrous layer, the annulus fibrosis
Related parts:	vertebrae, spinal ligaments

The Sacrum and Coccyx

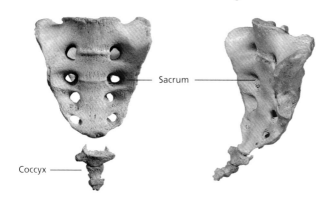

Sacrum

Coccyx

Inner surface of sacrum

Side view of sacrum

At the tail end of the vertebral column are two bony structures called the sacrum and coccyx. The sacrum is shaped like an upside-down triangle and consists of five sacral vertebrae that become fused together in early adulthood. It has several functions: it attaches the vertebral column to the bones of the pelvic girdle, supporting the body's weight; it protects the pelvic organs, such as the bladder; and it provides an attachment for the muscles that move the thigh. The coccyx is secured to the base of the sacrum and is the remains of the tail seen in our primate relatives. It consists of a small pyramid-shaped bone formed from four fused vertebrae and allows the attachment of the ligaments and muscles that form the anal sphincter.

Body system:	musculoskeletal system
Location:	at the base of the vertebral column
Function:	attach vertebral column to pelvic girdle, helping to support body's weight; protect pelvic organs and provide attachment for muscles
Components:	nine fused vertebrae
Related parts:	vertebral column, pelvic girdle, various muscles

Sacral Plexus

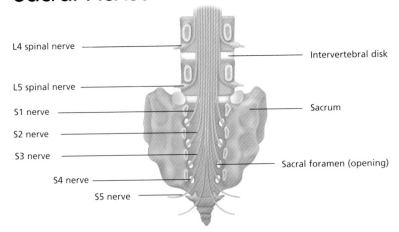

L4 spinal nerve

Intervertebral disk

L5 spinal nerve

S1 nerve

Sacrum

S2 nerve

S3 nerve

Sacral foramen (opening)

S4 nerve

S5 nerve

The sensory and motor nerve supply to and from the pelvis and legs is derived from a network of nerve roots called the sacral plexus, which lies in front of the sacrum. The nerves in the sacral plexus originate in the fourth and fifth lumbar nerve roots and the sacral nerve roots. At the sacral plexus, these nerve roots exchange nerve fibers and reform into major nerves. These include the superior and inferior gluteal nerves that supply the buttocks, and the sciatic nerve, which supplies the muscles of the leg. The parasympathetic splanchnic nerves (S1, S2, S3) regulate urination and defecation by controlling the internal sphincters, and they also play a role in erection by dilating the arteries in the penis.

Body system :	central nervous system
Location:	the rear wall of the pelvic cavity in front of the sacrum
Function:	network of nerves that supply the buttocks, legs, and genitals; also controls urination, defecation, and penile erection
Components:	sacral nerve roots S1–S5, lumbar nerve roots L4–L5
Related parts:	spinal cord, numerous nerves

The Spinal Cord

The spinal cord is a two-way communication pathway between the brain and the body. Thirty-one pairs of spinal nerves leave the spinal cord via openings in the spine called foramina, enabling signals to travel downward to control the body's functions, and upward to return information to the brain. The cord is a slightly flattened, cylindrical structure about 16½ in (42 cm) in length. It begins as a continuation of the medulla, the lower part of the brain stem, and runs the length of the back in the vertebral canal, protected by the bony vertebrae. The cord itself terminates at the conus medullaris, from where nerves continue to pass down as the cauda equina. Like the brain, the spinal cord is covered and protected by three meninges (membranes)—the dura, arachnoid, and pia maters—and is surrounded by cerebrospinal fluid.

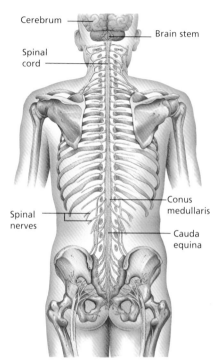

Cerebrum

Brain stem

Spinal cord

Conus medullaris

Spinal nerves

Cauda equina

Body system:	central nervous system
Location:	extends from the medulla in the brain stem to the lumbar vertebrae
Function:	two-way pathway between brain and body, allowing vital nerve impulses to travel between the two
Components:	gray matter and white matter containing nerve cells
Related parts:	vertebral column, brain, many other parts of the body

Spinal Cord Tracts

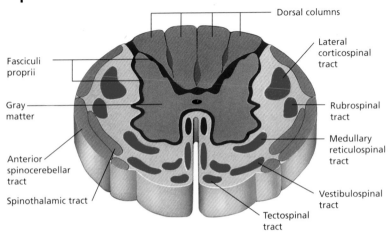

In cross-section, the spinal cord typically has a butterfly-shaped center of gray matter surrounded by white matter. Within the white matter are numerous paired tracts, or collections of nerve axons. All have the same origin, destination, and function. Ascending tracts (shown here as blue) carry sensory information concerning touch, pressure, pain, and temperature from the body to the brain. The descending tracts (red) carry signals from the brain to the body and are particularly involved with the control of movement. Extrapyramidal tracts (purple) have nerve fibers that pass in both directions and are concerned with the control of balance and coordination, posture, and muscle tone.

Body system:	central nervous system
Location:	within the white matter of the spinal cord
Function:	carry sensory information from the body to the brain, carry signals from the brain regarding movement, balance, and posture
Components:	ascending tracts, descending tracts, extrapyramidal tracts
Related parts:	vertebral column, brain, many of the structures in the body

Spinal Nerves

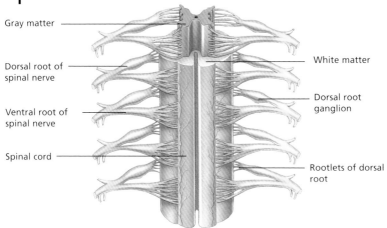

Gray matter

Dorsal root of
spinal nerve

Ventral root of
spinal nerve

Spinal cord

White matter

Dorsal root
ganglion

Rootlets of dorsal
root

There are 31 pairs of spinal nerves, arranged on each side of the spinal cord along its length. The pairs are grouped by region: eight cervical, 12 thoracic, five lumbar, five sacral, and one coccygeal. Each spinal nerve has two roots. The anterior, or ventral, root contains the axons of motor nerves, which send impulses to control muscle movement. The posterior, or dorsal, root contains the axons of sensory nerves, which send sensory information from the body into the spinal cord on its way to the brain. Each root is formed by a series of small rootlets that attach it to the cord. The portion of the spinal cord that provides the rootlets for one dorsal root is referred to as a segment.

Body system:	central nervous system
Location:	arranged on either side of the spinal cord along its length
Function:	carry sensory information from the body to the brain; carry motor impulses from the brain to the muscles
Components:	8 cervical, 12 thoracic, 5 lumbar, 5 sacral, 1 coccygeal
Related parts:	brain, spinal cord, many structures in the body

103

Spinal Meninges

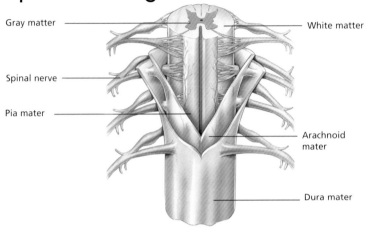

Gray matter

White matter

Spinal nerve

Pia mater

Arachnoid mater

Dura mater

The bones of the vertebral column provide the main protection for the spinal cord. However, like the brain, the cord has additional protection from three membranes—the meninges—that continue down from inside the skull. The dura mater is the tough outer membrane and is separated from the vertebrae by the epidural space, which contains fatty tissue and veins. The middle membrane is the delicate arachnoid mater, and between this and the dura mater there is a thin film of cerebrospinal fluid (CSF). The inner membrane, the pia mater, lies against the spinal cord and is richly supplied with fine blood vessels. Between the arachnoid mater and the pia mater is the subarachnoid space; this contains CSF to "cushion" the cord.

Body system:	central nervous system
Location:	surround the spinal cord
Function:	provide protection for the cord; cerebrospinal fluid acts as a "cushion" and helps remove chemical waste
Components:	dura mater, arachnoid mater, pia mater
Related parts:	brain, spinal cord

Superficial Back Muscles

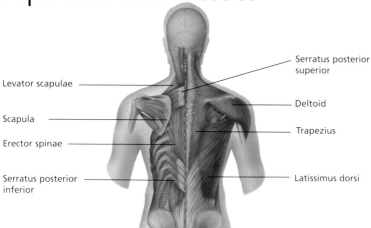

Levator scapulae

Scapula

Erector spinae

Serratus posterior inferior

Serratus posterior superior

Deltoid

Trapezius

Latissimus dorsi

The superficial muscles of the back act with other muscles to move the neck, back, shoulders, and upper arms. The trapezius is a large fan-shaped muscle, the top edge of which forms the visible slope from the neck to the shoulder. It attaches to the base of the skull and helps hold up and rotate the head and brace the shoulders back. The largest, most powerful muscle in the back, the latissimus dorsi, is attached to the spine and runs down to the pelvis. This muscle allows the arm to be pulled back in line with the trunk, even against great force. The rotator cuff is a group of smaller muscles that runs between the scapula and the head of the humerus; together they hold the humerus tightly into the shoulder joint.

Body system:	musculoskeletal system
Location:	between the neck and the pelvis
Function:	enable movement of the neck, back, and arms; one group of muscles moves the rib cage up during breathing
Components:	trapezius, latissimus dorsi, levator scapulae, rotator cuff, erector spinae, serratus posterior
Related parts:	skull, vertebral column, ribs, humerus, shoulders

Deep Back Muscles

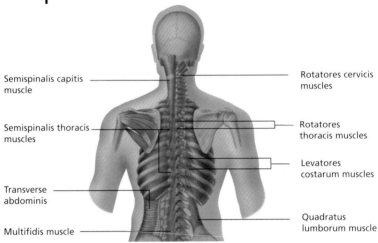

Semispinalis capitis muscle

Semispinalis thoracis muscles

Transverse abdominis

Multifidis muscle

Rotatores cervicis muscles

Rotatores thoracis muscles

Levatores costarum muscles

Quadratus lumborum muscle

Beneath the superficial muscle layer of the back are deeper muscles, which are attached to the vertebrae, ribs, base of the skull, and pelvis. These muscles are responsible for the smooth movements of the spine and are arranged in layers. The most deeply located muscles are very short and run obliquely from each vertebra to the one above. Over these lie muscles that are longer and run between several vertebrae and nearby ribs. More superficially, the muscles become even longer and some are attached to the pelvic bones and the occiput (base of the skull). Together these muscles act in conjunction to maintain the spine in an S-shaped curve, enabling its fluid movements and upright posture.

Body system:	musculoskeletal system
Location:	between the neck and the pelvis
Function:	allow smooth movements of spine, maintain erect posture
Components:	semispinalis capitis muscle, semispinalis thoracis muscles, multifidis muscle, rotatores cervicis muscles, rotatores thoracis muscles, levatores costarum muscles, quadratus lumborum muscle
Related parts:	skull, vertebral column, ribs, pelvis

The Clavicle

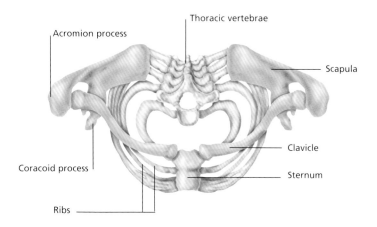

Acromion process

Thoracic vertebrae

Scapula

Coracoid process

Clavicle

Sternum

Ribs

The arms are connected to the skeleton at the pectoral girdle, which is made up of the clavicle (collarbone) and the scapula (shoulder blade). The clavicle is a slender S-shaped bone that lies horizontally at the upper border of the chest and acts as a strut to hold the upper limbs away from the thorax. The front and upper surfaces of the clavicle are mostly smooth, while the under-surfaces are ridged and grooved by the attachments of ligaments and muscles. The inner end of the clavicle has a large oval facet for connecting with the sternum at the sternoclavicular joint. A smaller facet lies at the other end where the clavicle articulates with the acromion process (a bony prominence on the scapula) at the acromioclavicular joint.

Body system:	musculoskeletal system
Location:	lies horizontally across the upper border of the chest
Function:	provides an attachment for muscles and ligaments; holds upper limbs away from chest cavity
Components:	two bones, each connected to sternum and scapula
Related parts:	humerus, sternum, scapula, ribs, muscles

The Sternoclavicular Joint

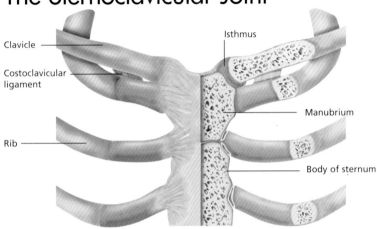

Clavicle

Costoclavicular ligament

Rib

Isthmus

Manubrium

Body of sternum

The sternoclavicular joint is the only bony connection between the pectoral girdle and the rest of the skeleton. The joint can be felt under the skin as the sternal end of the clavicle is fairly large and extends above the top of the manubrium (upper section of the sternum), both sides together forming the familiar sternal notch at the base of the neck. Inside the joint is an articular disk made of fibrocartilage, which improves the fit of the bones and keeps them stable. The joint is further stabilized by the costoclavicular ligament, which anchors its underside to the first rib. Only a small degree of movement is possible at the sternoclavicular joint, although the outer end of the clavicle can move upward, as when shrugging.

Body system :	musculoskeletal system
Location:	between the clavicle and the sternum
Function:	stablizes the junction of the clavicle and sternum
Components:	articular disk, costoclavicular ligament
Related parts:	ribs

The Scapula

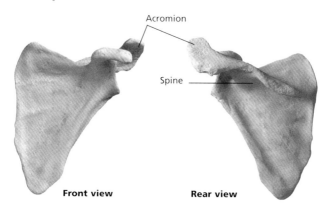

Acromion

Spine

Front view　　**Rear view**

A long with the clavicle (collarbone), the scapula (shoulder blade) forms the bony structure called the pectoral girdle. It is a thin, flat, triangular-shaped bone that has two surfaces: anterior (front) and posterior (back). The anterior or costal surface lies against the ribs at the back of the chest and is concave—the hollow area, which is known as the subscapular fossa, provides a large surface area for the attachment of muscles. The posterior surface is divided by a prominent spine, which runs horizontally across the back of the scapula and can be felt under the skin of the upper back. This spine is continuous with the bony projection called the acromion, which forms the tip of the shoulder and connects with the clavicle.

Body system:	musculoskeletal system
Location:	lies against the back of the ribcage between ribs 2 and 7
Function:	provides an attachment for muscles and ligaments; holds upper limbs away from chest cavity
Components:	three-sided flat bone, spine, acromion, subscapular fossa
Related parts:	humerus, clavicle, ribs

The Rib Cage

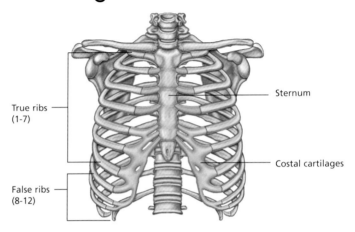

True ribs
(1-7)

False ribs
(8-12)

Sternum

Costal cartilages

The rib cage protects the vital organs in the chest, as well as provides sites for the attachment of muscles in the back, chest, and shoulder. It consists of 12 paired ribs, the costal cartilages, and the sternum. The rib cage is supported at the back by the 12 thoracic vertebrae and each of the ribs is attached to the corresponding numbered vertebra. The ribs then curve down and around the chest toward the front of the body. The ribs can be divided into two groups—true ribs and false ribs—according to their front site of attachment. The first seven pairs are true ribs and attach directly to the sternum via individual costal cartilages. Rib pairs eight to ten attach indirectly via fused cartilages and pairs 11 and 12 are "floating" ribs.

Body system:	musculoskeletal system
Location:	beneath the neck, encompassing the chest
Function:	protects vital chest organs, provides an attachment site for various muscles
Components:	ribs, costal cartilages, sternum
Related parts:	vertebral column, pectoral girdle, lungs

The Sternum

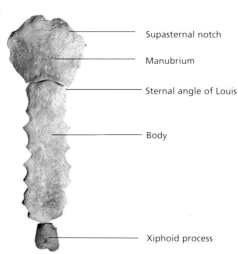

Supasternal notch

Manubrium

Sternal angle of Louis

Body

Xiphoid process

At the front of the rib cage is a long, flat bone called the sternum, or breastbone. It has three parts: the manubrium, the body, and the xiphoid process. The manubrium forms the upper part of the sternum and is in the shape of a rough triangle with a prominent and palpable notch in its center. The body forms the greater length of the sternum and is at a slight angle to the manubrium, forming a joint that allows for movements during respiration. At the lower end of the sternum is the xiphoid process, a small pointed bone that projects downward and slightly backward and provides an attachment for some abdominal muscles. In young people, it may be cartilaginous, but it usually changes to bone by 40–50 years of age.

Body system:	musculoskeletal system
Location:	at the front of the rib cage
Function:	forms the front of the rib cage, and protects the vital organs in the chest; provides an attachment for some abdominal muscles
Components:	manubrium, body, xiphoid process
Related parts:	ribs, pectoral girdle

Costal Cartilages

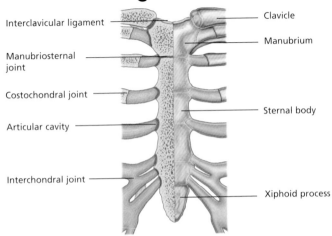

Interclavicular ligament

Manubriosternal joint

Costochondral joint

Articular cavity

Interchondral joint

Clavicle

Manubrium

Sternal body

Xiphoid process

The ribs are connected to the sternum (breastbone) by the costal cartilages. These flexible and resilient structures are made of "hyaline" cartilage, which is tough but elastic, and so their presence contributes to the mobility of the thoracic wall. During the process of breathing, the costal cartilages stretch and twist to allow the rib cage to lift and expand as air is inhaled into the lungs, and afterward to "spring back" to regain their shape and position. The first seven costal cartilages attach directly to the sternum, the next three attach to the costal cartilage directly above them, while the last two are really just caps of cartilage on the ends of the rib shafts, which attach only to the soft tissues of the lateral abdominal wall.

Body system :	musculoskeletal system
Location:	between the ribs and the sternum
Function:	provide flexibility of movement for the ribcage
Components:	hyaline cartilage
Related parts:	ribs, sternum, lungs

Intercostal Muscles

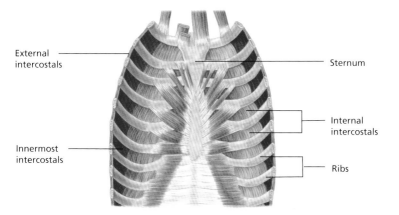

External intercostals

Sternum

Internal intercostals

Innermost intercostals

Ribs

The bony skeleton of the rib cage is sheathed in several layers of muscle, which include many of the powerful muscles of the upper limbs and back. The integral muscles of the rib cage, however, are only concerned with breathing. They help form the structure of the thoracic wall, enclosing the internal organs of the thorax. The intercostal muscles fill the 11 intercostal spaces between the ribs and lie in layers. The external intercostals are superficial, and their contraction acts to lift the ribs when breathing in. The internal intercostals also act to assist in breathing. The innermost intercostals are at the deepest level and are separated from the internal intercostals by connective tissue containing nerves and blood vessels.

Body system:	musculoskeletal system
Location:	between the ribs
Function:	enable rib cage to expand and contract during breathing
Components:	internal, external, and innermost intercostals
Related parts:	ribs, sternum, chest wall, accessory muscles, lungs

Movements of the Rib Cage

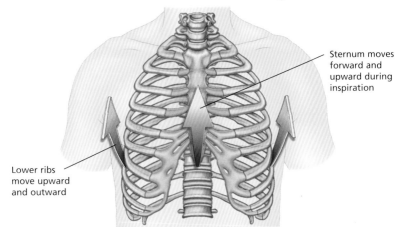

Sternum moves forward and upward during inspiration

Lower ribs move upward and outward

In order for inspiration (breathing in) to be possible, the pressure inside the lungs must be lower than the pressure in the atmosphere. This pressure difference is achieved by expanding the volume of the lungs (Boyle's law) and, as a result, air naturally flows into them to equalize the pressure. The principal structures involved in this process are the external intercostal muscles and the diaphragm, which contract, causing the rib cage to expand outward and upward. Expiration (breathing out) is a more passive process. The muscles used during inspiration relax, the ribs move downward and the diaphragm moves upward, returning the lungs to their resting size and forcing out the air within them.

Body system:	musculoskeletal system
Location:	beneath the neck, encompassing the chest
Function:	increase volume of lungs, reducing pressure, and causing air to be sucked in
Components:	intercostal muscles, diaphragm, rib cage
Related parts:	lungs

Accessory Muscles of Respiration

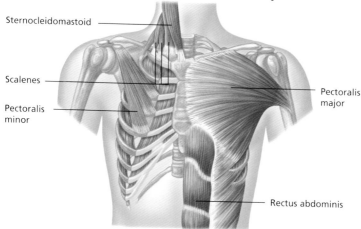

Sternocleidomastoid

Scalenes

Pectoralis minor

Pectoralis major

Rectus abdominis

There are situations when a much greater volume of air must enter the lungs (for example, during exercise) or when, due to lung disease, there is an increased resistance to the entry of air. At these times, the accessory muscles of respiration are brought into play. These are muscles with attachments both to the rib cage and to other parts of the upper skeleton, their normal function being to move the head, neck, or upper limbs. The powerful sternocleidomastoid muscles in the neck, for example, usually turn the head but can also be used in deep inspiration. The pectoralis muscles in the chest wall can help to pull the rib cage upward and outward and the rectus abdominus aids in forced expiration, such as coughing.

Body system:	musculoskeletal system
Location:	various locations in the upper body
Function:	aid breathing when the respiratory muscles are potentially inadequate
Components:	sternocleidomastoid, scalenes, pectoralis major and minor, rectus abdominis
Related parts:	rib cage, sternum, pectoral girdle, lungs

Arteries of the Thoracic Wall

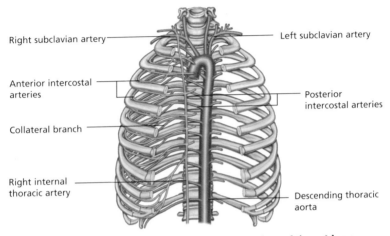

Right subclavian artery

Left subclavian artery

Anterior intercostal arteries

Posterior intercostal arteries

Collateral branch

Right internal thoracic artery

Descending thoracic aorta

The thoracic wall (the rib cage and surrounding muscles and tissues) has a plentiful blood supply carried by the intercostal arteries, which run along the spaces between the ribs. The arteries create a network of blood vessels that encircles the thoracic wall and supplies all its structures. Each intercostal space has within it a posterior intercostal artery, which originates near the spine, and two anterior intercostal arteries, which arise from the internal thoracic arteries running vertically down either side of the sternum. Each posterior artery has a dorsal branch, which travels backward to supply the spine and back muscles, and a small collateral artery, which travels along the upper surface of the rib below.

Body system :	cardiovascular system
Location:	encircling the internal thoracic wall
Function:	supply blood rich in oxygen and nutrients to the rib cage and surrounding muscles and tissues
Components:	posterior and anterior intercostal arteries, internal thoracic arteries
Related parts:	aorta, subclavian artery

Veins of the Thoracic Wall

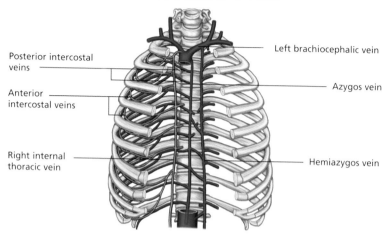

Posterior intercostal veins

Left brachiocephalic vein

Anterior intercostal veins

Azygos vein

Right internal thoracic vein

Hemiazygos vein

The intercostal veins run alongside the intercostal arteries in the spaces between the ribs and drain deoxygenated blood from the rib cage and surrounding muscles and tissues. There are 11 posterior intercostal veins and one subcostal vein (lying beneath the twelfth rib) on each side of the sternum (breastbone) which, like the arteries, communicate with the corresponding anterior intercostal vessels to form a network around the rib cage. The posterior veins drain blood back to the azygos venous system, which lies in front of the spine at the back of the thoracic wall. Like the arteries in the same position, the anterior veins drain into the internal thoracic veins, which run alongside the internal thoracic arteries.

Body system:	cardiovascular system
Location:	encircling the internal thoracic wall
Function:	drain deoxygenated blood from the rib cage and surrounding muscles and tissues
Components:	posterior and anterior intercostal veins, internal thoracic veins
Related parts:	azygos vein, superior vena cava, brachiocephalic vein

Intercostal Nerves

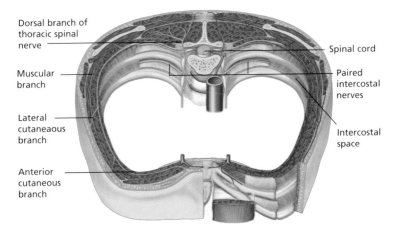

Dorsal branch of thoracic spinal nerve

Spinal cord

Muscular branch

Paired intercostal nerves

Lateral cutaneaous branch

Intercostal space

Anterior cutaneous branch

There are 12 pairs of spinal nerves that arise from the thoracic part of the spinal cord corresponding to the thoracic vertebrae. These give off posterior branches that supply the skin and muscles of the back. The anterior branches of the thoracic spinal nerves become the intercostal nerves (except the twelfth, which becomes the subcostal nerve). Each intercostal nerve travels together with its corresponding artery and vein protected in the costal groove along the lower edge of each rib. A typical intercostal nerve has both sensory and motor fibers and has four branches: collateral, lateral cutaneous, anterior cutaneous, and muscular. Between them, these branches supply the muscles and some tissues within the thoracic wall.

Body system:	nervous system
Location:	arise from spinal cord and run around thoracic wall
Function:	provide motor and sensory nerve supply to muscles and tissues of thoracic wall
Components:	collateral, lateral cutaneous, anterior cutaneous, muscular branches
Related parts:	spinal cord, other structures in thoracic wall

The Male Breast

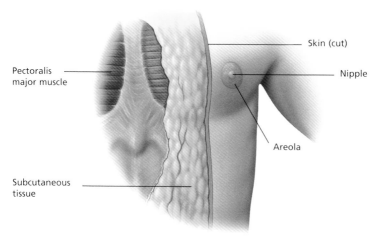

Skin (cut)

Nipple

Pectoralis
major muscle

Areola

Subcutaneous
tissue

Men and women both have breast tissue but the breasts are usually only well developed structures in women. Although the male breast appears very different to the female breast, its underlying structure is similar. Both are formed of fat and glandular tissue and have a central nipple containing ducts, surrounded by a pigmented area—the areola. The difference lies in the fact that during puberty the hormones that stimulate the enlargement and further development of breast tissue in females are not present in males. The male breast overlies the pectoralis major, a large fan-shaped muscle at the front of the chest, and if this muscle is well developed, the male breasts can become prominent.

Body system:	integumentary system
Location:	overlying the pectoralis major at the front of the chest
Function:	in males, the breasts have no obvious function
Components:	fat and undeveloped glandular tissue
Related parts:	pectoralis major muscle

The Female Breast

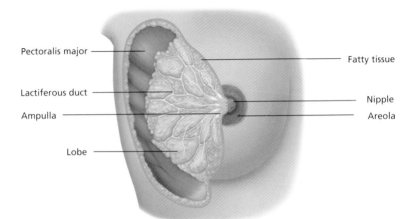

Pectoralis major

Fatty tissue

Lactiferous duct

Nipple

Ampulla

Areola

Lobe

The female breast, or mammary gland, is hemispherical in shape and extends from the level of the second rib above and the sixth rib below. In addition, there may be an extension of breast tissue toward the armpit, known as the axillary tail. Just below the center of each breast is an area of pigmented skin called the areola, which surrounds the nipple. Within each breast there are about 15–20 lobes that contain smaller structures called lobules—the glandular tissue that secretes milk after childbirth. A system of lactiferous ducts passes the milk to the nipple, where the ducts open onto the surface of the skin. In nonpregnant women, the breast tissue is largely composed of fat and the glandular structure remains undeveloped.

Body system:	reproductive system
Location:	at the front of the chest wall
Function:	produce breast milk to nourish a newborn baby
Components:	fat, glandular tissue, nipple, areola
Related parts:	pectoralis major muscle

Blood Supply to the Breast

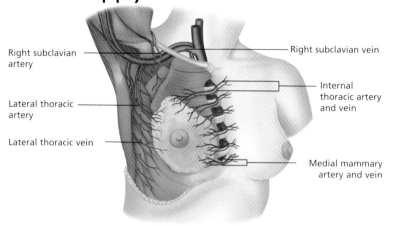

Right subclavian artery

Lateral thoracic artery

Lateral thoracic vein

Right subclavian vein

Internal thoracic artery and vein

Medial mammary artery and vein

The arterial blood supply to the breast comes from a number of sources; these include the internal thoracic artery, which runs down the length of the front of the chest and gives off branches that enter the breast tissue, and the lateral thoracic artery, which supplies the outer part of the breast. A network of superficial veins underlies the skin of the breast, especially in the region of the areola, and these veins often become very prominent during pregnancy. The blood collected in these veins drains in various directions, following a similar pattern to the arterial supply, and travels via the internal thoracic veins and the posterior intercostal veins to the large veins that return blood to the heart.

Body system:	cardiovascular system
Location:	within the chest wall and breast tissue
Function:	delivers blood rich in oxygen and nutrients to the breast tissue and takes deoxygenated blood back to the heart
Components:	lateral thoracic artery and vein, mammary artery and vein, internal thoracic artery and vein, subclavian artery and vein
Related parts:	other nearby blood vessels, heart

Lymphatic Drainage of the Breast

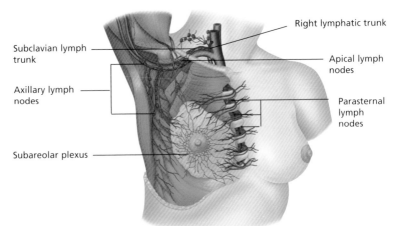

Right lymphatic trunk

Subclavian lymph trunk

Apical lymph nodes

Axillary lymph nodes

Parasternal lymph nodes

Subareolar plexus

Lymph, the fluid that leaks out of blood vessels into the spaces between the cells, is returned to the blood circulation by the lymphatic system. Lymph nodes located at intervals in the system act as filters to remove bacteria, cells, and other particles. Lymph drains from the nipple, areola, and mammary gland lobules into a network of small lymphatic vessels, the subareolar lymphatic plexus. About 75 percent of the lymph from the subareolar plexus drains to lymph nodes in the axilla (armpit), and from there to the subclavian lymph trunk. The remaining lymph, mainly from the inner quadrants of the breast is carried to parasternal lymph nodes, which lie toward the midline of the front of the chest.

Body system:	lymphatic system
Location:	within the breast tissue, axilla, chest wall
Function:	drains excess fluid from around the cells in breast tissue; filters fluid to remove bacteria and particles
Components:	subareolar plexus, axillary nodes, interpectoral nodes, subclavian lymph trunk, parasternal nodes
Related parts:	blood vessels

Abdominal Surface of the Diaphragm

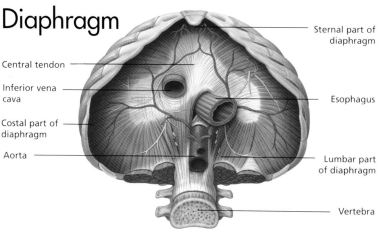

Central tendon

Inferior vena cava

Costal part of diaphragm

Aorta

Sternal part of diaphragm

Esophagus

Lumbar part of diaphragm

Vertebra

The diaphragm is a powerful sheet of muscle that separates the thorax (chest cavity) from the abdomen. It is the main muscle involved with breathing and has several openings to allow the passage of important structures, such as the esophagus and major blood vessels. The muscle tissue of the diaphragm arises from three areas of the chest wall, which give rise to separately named parts: the sternal, costal, and lumbar or vertebral parts. These parts merge to form a continuous sheet that converges on a central tendon, which acts as a site of muscular attachment. The tendon has a characteristic three-leaved shape and, unlike other tendons, does not have any attachment to bone.

Body system:	musculoskeletal system
Location:	separates the thorax from the abdomen
Function:	main muscle of breathing; its contraction expands the chest cavity causing air to enter the lungs
Components:	sternal, costal, and lumbar parts, central tendon
Related parts:	structures in the chest cavity and abdomen, major blood vessels, vertebrae, ribs

Thoracic Surface of the Diaphragm

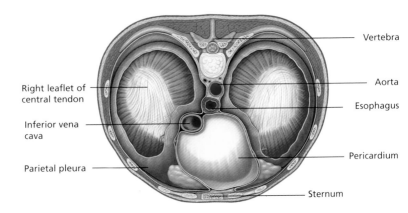

Vertebra

Aorta

Esophagus

Pericardium

Sternum

Right leaflet of central tendon

Inferior vena cava

Parietal pleura

The upper aspect of the diaphragm is convex and forms the floor of the chest cavity. It is perforated by major vessels and structures, which must pass through the muscle sheet in order to reach the abdomen. The three largest openings in the diaphragm are the caval aperture, through which the inferior vena cava passes, the esophageal aperture, which allows the esophagus to enter the abdomen, and the aortic aperture next to the spine, through which the aorta passes. The central part of the surface of the diaphragm is covered by the pericardium, the membrane that surrounds the heart. To either side the surface is lined with the diaphragmatic part of the parietal pleura (the thin membrane that lines the pleural cavity).

Body system :	musculoskeletal system
Location:	separates the thorax from the abdomen
Function:	main muscle of breathing; its contraction expands the chest cavity causing air to enter the lungs
Components:	apertures, pericardium, parietal pleura, muscle, tendon
Related parts:	structures in the chest cavity and abdomen, major blood vessels, vertebrae, ribs

Nerve Supply of the Diaphragm

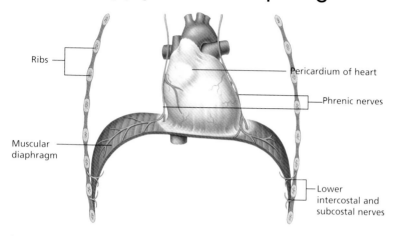

Ribs

Pericardium of heart

Phrenic nerves

Muscular diaphragm

Lower intercostal and subcostal nerves

The motor nerve supply of the diaphragm (which causes contraction of the muscle fibers) comes entirely from the phrenic nerves, spread out across the surface of the muscle. These paired nerves originate from the cervical plexus (an interlacing nerve network) in the neck and approach the diaphragm along the fibrous pericardium, the outer layer of the heart. The phrenic nerves also provide a sensory nerve supply to the central part of the diaphragm, detecting pain and giving information on position. Irritation of these nerves can cause intermittent spasm of the diaphragm, commonly known as hiccups. The edges of the diaphragm receive a sensory nerve supply from the lower intercostal and subcostal nerves.

Body system:	nervous system
Location:	phrenic nerves run from cervical plexus to diaphragm
Function:	provide nerve supply to diaphragm causing the muscle fibers to contract
Components:	phrenic nerves, intercostal nerves, subcostal nerves
Related parts:	spinal cord, heart

The Esophagus

The esophagus is the flexible, muscular tube that provides a passageway for food from the mouth to the stomach. When the esophagus is empty, the tube is "collapsed" and the inner lining lies in folds that fill the lumen, or central space. As food is swallowed and passes down, it distends the lining and the esophageal walls and is carried down the esophagus by waves of muscular contraction known as peristalsis. The esophagus pierces the diaphragm at the esophageal aperture and enters the stomach at the cardiac orifice. In cross-section, the esophagus has four layers: the innermost mucosal layer, which is resistant to abrasion by food; the submucosal layer, which contains glands that secrete mucus to aid the passage of food; the muscle layer; and the adventitia, which is a covering layer of fibrous connective tissue.

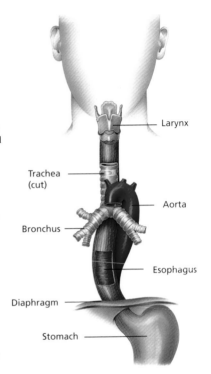

Larynx

Trachea (cut)

Aorta

Bronchus

Esophagus

Diaphragm

Stomach

Body system:	digestive system
Location:	runs from the throat to the stomach
Function:	provides passageway for food from the mouth to the stomach; propels food downward by muscular contractions
Components:	mucosal layer, submucosal layer, muscle, adventitia
Related parts:	mouth, pharynx, stomach, diaphragm, trachea, aorta

Nerves of the Esophagus

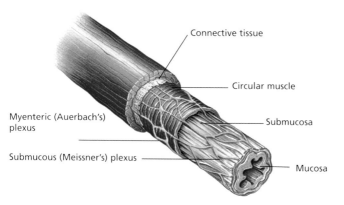

Connective tissue

Circular muscle

Myenteric (Auerbach's) plexus

Submucosa

Submucous (Meissner's) plexus

Mucosa

In common with the rest of the gastrointestinal tract, the esophagus has its own intrinsic nerve supply, which allows it to contract and relax during the process of peristalsis (waves of muscular contraction) without any external stimulation. This intrinsic nerve supply derives from two main nerve plexuses (networks of nerves) within the walls of the esophagus, the submucous (Meissner's) plexus, and the myenteric (Auerbach's) plexus. These connect with each other, and together regulate the glandular secretion and movements of the esophagus. The functioning of the intrinsic system can be modified by the autonomic nervous system, which regulates the body's internal environment.

Body system:	nervous system
Location:	within the walls of the esophagus
Function:	stimulates the muscular walls of the esophagus to contract and relax, propelling food down toward the stomach
Components:	submucous plexus and myenteric plexus
Related parts:	sympathetic trunk and vagus nerve

The Lungs

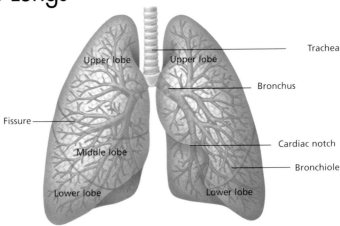

Trachea

Upper lobe

Upper lobe

Bronchus

Fissure

Cardiac notch

Middle lobe

Bronchiole

Lower lobe

Lower lobe

The paired lungs are soft, spongy, cone-shaped organs of respiration that occupy the thoracic cavity and lie on either side of the heart. Each lung is enclosed in a membranous bag, the pleural sac, and has an apex, which projects up into the base of the neck behind the clavicle (collarbone), a base that rests on the upper surface of the diaphragm and a concave mediastinal surface. The lungs are divided into sections, know as lobes, by deep fissures lined with pleural membrane. The right lung consists of three lobes, while the left lung, which is slightly smaller (due to the position of the heart) has only two. Each lobe is independent of the others, receiving inhaled air via its own lobar bronchus (airway).

Body system:	respiratory system
Location:	in the thoracic cavity on either side of the heart
Function:	provide the body with a continuous supply of oxygen and the means to dispose of waste carbon dioxide
Components:	alveoli, bronchioles, bronchi, pleurae
Related parts:	cardiovascular system, upper airways

The Pleura

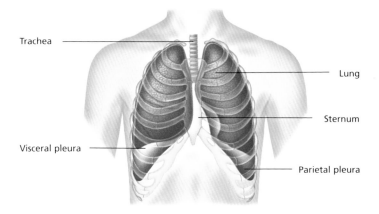

Trachea

Lung

Sternum

Visceral pleura

Parietal pleura

Each lung is covered with a thin membrane known as the pleura, which lines both the outer surface of the lung (visceral pleura) and the inner surface of the chest wall (parietal pleura). The visceral pleura covers the lung surface, dipping down into the fissures between the lobes. The parietal pleura extends on from the visceral pleura at the root of the lung and continues to cover the thoracic wall and the surfaces of structures in the cavity. The parietal pleura is itself divided into areas: the costal pleura lines the inside of the rib cage; the mediastinal pleura covers the mediastinum; the diaphragmatic pleura lines the upper surface of the diaphragm; and the cervical pleura covers the tip of the lung as it projects into the neck.

Body system:	respiratory system
Location:	covering the lungs
Function:	provides a smooth slippery surface, enabling the lung to slide easily against the rib cage during breathing
Components:	parietal and visceral pleurae
Related parts:	lungs, rib cage, diaphragm, mediastinum

Pleural Cavity and Recesses

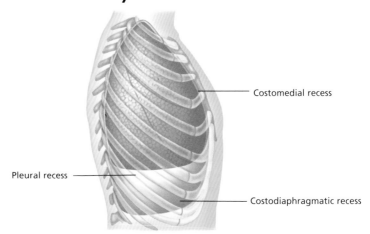

Costomedial recess

Pleural recess

Costodiaphragmatic recess

The pleural cavity, which lies between the visceral and parietal layers of pleura, is a narrow space filled with a small amount of pleural fluid. The fluid lubricates the movement of the lung within the thoracic cavity and also acts to provide a tight seal, holding the lung against the thoracic wall and diaphragm by surface tension. It is this seal that forces the elastic tissue of the lung to expand when the diaphragm contracts and the ribcage lifts during breathing in. During quiet breathing, the lungs do not completely fill the pleural sacs within which they lie. There is room for expansion in the pleural recesses—areas where the sacs are empty—but the lungs only expand into these recesses during deep breathing.

Body system:	respiratory system
Location:	between the visceral and parietal pleurae
Function:	pleural fluid lubricates pleurae and provides a tight seal; recesses allow lungs to fill to maximum volume during deep breathing
Components:	pleural fluid, costomedial recess, costodiaphragmatic recess, pleural recess
Related parts:	lungs, ribcage

The Trachea and Bronchi

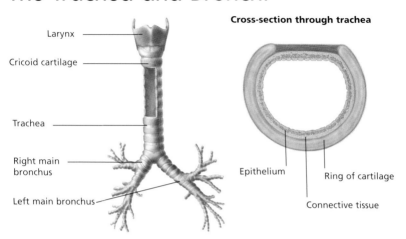

Larynx

Cricoid cartilage

Trachea

Right main bronchus

Left main bronchus

Cross-section through trachea

Epithelium

Ring of cartilage

Connective tissue

As a breath is taken, air enters the body and passes through the larynx to the trachea (main airway). The trachea extends from the cricoid cartilage just below the larynx to the chest where it divides into the two main bronchi that lead to the lungs. The trachea is composed of strong fibroelastic tissue, within which are embedded a series of incomplete rings of hyaline cartilage—the tracheal cartilages. The back of the trachea has no cartilage and lies in contact with the esophagus. On entering the lung, the two main bronchi divide repeatedly to form smaller bronchi; at each division the airways become smaller. The bronchi have a similar structure to the trachea, but they also have muscle fibers in their walls.

Body system:	respiratory system
Location:	trachea runs from the cricoid cartilage to the level of the sternal angle, bronchi branch from the trachea
Function:	take air to the lungs; lining of airways secretes mucus to trap tiny particles
Components:	trachea, right and left main bronchi, numerous smaller bronchi
Related parts:	mouth, nose, larynx, lungs

Bronchioles and Alveoli

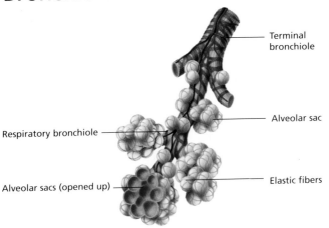

Terminal bronchiole

Alveolar sac

Respiratory bronchiole

Elastic fibers

Alveolar sacs (opened up)

When the airways have an internal diameter of less than ¾₄ in (1 mm), they are known as bronchioles. Bronchioles differ from bronchi in that they have no cartilage in their walls, nor any mucus-secreting cells in their lining. Further divisions lead to the formation of terminal bronchioles, which in turn become respiratory bronchioles, the smallest of all the air passages. There are millions of alveoli in each lung (together giving a surface area of about 1,500 ft²/140 m²) clustered around the respiratory bronchioles like grapes. These tiny, hollow sacs have thin walls, through which oxygen diffuses from the air into the pulmonary bloodstream and carbon dioxide passes out of the blood into the lungs.

Body system:	respiratory system
Location:	in the lungs
Function:	bronchioles channel air toward the alveoli, the functioning parts of the lung in which gaseous exchange takes place
Components:	terminal and respiratory bronchioles, alveolar ducts, alveolar sacs
Related parts:	bronchi, trachea, blood vessels

Pulmonary Circulation

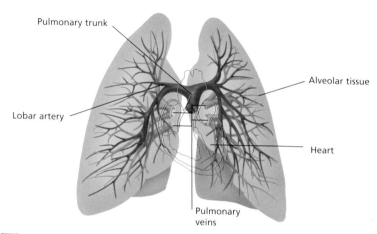

Pulmonary trunk

Alveolar tissue

Lobar artery

Heart

Pulmonary veins

The primary function of the lungs is to reoxygenate the blood used by the tissues of the body and to remove waste carbon dioxide. A large artery known as the pulmonary trunk carries deoxygenated blood from the heart's right ventricle to the lungs. The pulmonary trunk divides into two smaller branches, the right and left pulmonary arteries, which enter the lungs at the same point as the two main bronchi (airways). Within the lungs the arteries divide to supply each lobe of their respective lung, two on the left and three on the right, and terminate in a network of capillaries where gaseous exchange takes place. Freshly oxygenated blood returns to the heart through a system of pulmonary veins that run alongside the arteries.

Body system:	respiratory system
Location:	within the lungs
Function:	deliver deoxygenated blood to the lungs for reoxygenation and removal of waste carbon dioxide
Components:	pulmonary trunk, pulmonary arteries, lobar arteries, pulmonary veins
Related parts:	other nearby blood vessels, heart

Alveolar Capillary Plexus

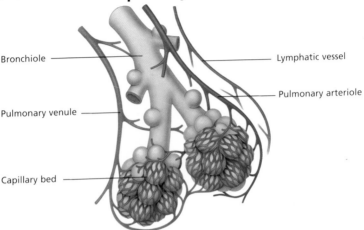

Bronchiole

Lymphatic vessel

Pulmonary arteriole

Pulmonary venule

Capillary bed

Within the lung, repeated division of the pulmonary arteries ultimately results in a network (plexus) of tiny blood vessels (capillaries) that surround each of the millions of alveolar sacs. The walls of the capillaries are extremely thin, which allows the blood within them to come into close contact with the walls of the alveoli. Oxygen diffuses into the pulmonary blood from the lungs and carbon dioxide passes from the blood back into the lungs to be exhaled (breathed out). The newly oxygenated blood is collected in tiny veins called venules, which drain into each capillary plexus and ultimately join to form the pulmonary veins. These complete the pulmonary circulation by returning blood to the heart.

Body system:	cardiovascular system
Location:	surrounding the alveoli
Function:	enable gaseous exchange to take place between the blood and the lung tissue
Components:	pulmonary arterioles, pulmonary venules, capillary bed
Related parts:	lungs, circulatory system

Lymphatics of the Lung

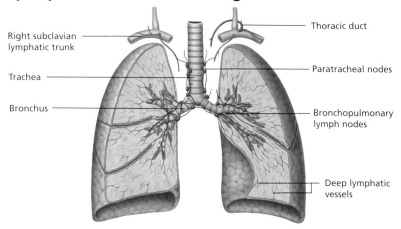

Right subclavian lymphatic trunk

Trachea

Bronchus

Thoracic duct

Paratracheal nodes

Bronchopulmonary lymph nodes

Deep lymphatic vessels

The lymphatic drainage of the lungs originates in two main networks or plexuses. The superficial plexus is a network of fine lymphatic vessels that extends over the surface of the lung just beneath the visceral pleura and drains lymph from the lung toward the bronchi and trachea, where the main groups of lymph nodes are found. The lymphatic vessels of the deep lymphatic plexus drain the deeper tissues of the lungs and originate in the connective tissue surrounding the small airways (bronchi and bronchioles). There are also small lymphatic vessels within the lining of the larger airways. Lymph nodes are scattered around the main airways and filter the lymph, playing an important role in the prevention of infection.

Body system:	lymphatic system
Location:	within and around the lung tissue
Function:	drain excess fluid from around the cells of the body and return it to the circulation; filter lymph to remove particles and bacteria
Components:	superficial and deep lymphatic plexus, lymph nodes
Related parts:	brachiocephalic vein, subclavian veins, respiratory system

The Mediastinum

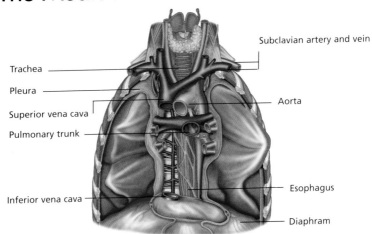

Subclavian artery and vein

Trachea

Pleura

Superior vena cava

Pulmonary trunk

Aorta

Esophagus

Inferior vena cava

Diaphram

The mediastinum is the central cavity in the chest that contains the heart and other vital structures. The cavity extends from the base of the neck to the diaphragm and from the sternum (breastbone) to the spine, and it is bordered on either side by the mediastinal pleurae (the membranes that surround the lungs). As well as the heart and major blood vessels, contained within the mediastinum are the thymus gland, the trachea (main airway), the esophagus, and some important nerves, including the vagus and phrenic nerves. The structures within the mediastinum are held only loosely together by fatty connective tissue, which allows for movements within the thorax during breathing and changes in body posture.

Body system :	houses structures from many body systems
Location:	in the center of the thoracic cavity
Function:	contains vital structures, such as the heart, esophagus, and trachea
Components:	cavity bordered by lungs, sternum, and spine
Related parts:	thoracic organs

The Thymus

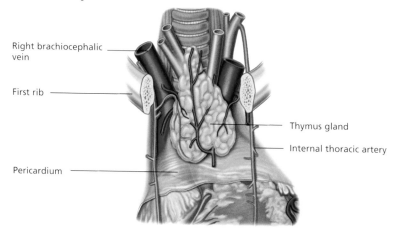

Right brachiocephalic vein

First rib

Thymus gland

Internal thoracic artery

Pericardium

The thymus gland is a pink, flattened structure that lies within the mediastinum (the central cavity in the thorax) and extends upward in front of the large blood vessels and trachea. It is usually formed of two lobes (bilobar), each divided into smaller lobules, which are enclosed within a capsule of connective tissue. The gland is very active during the early years of life, when it plays a major role in the development of the immune system. Hormones produced in the thymus gland help produce specialized white blood cells called T-lymphocytes, which fight infection. The thymus gland reaches its maximum size during puberty, but during adolescence the gland shrinks and by old age it has almost disappeared.

Body system:	endocrine and lymphatic systems
Location:	within the mediastinum behind the sternum (breastbone)
Function:	produces hormones that enable T-lymphocytes to mature; T-lymphocytes play a major role in fighting infection
Components:	two lobes, divided into numerous lobules
Related parts:	blood circulation, immune system

The Heart

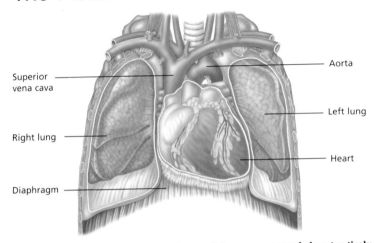

Superior vena cava

Aorta

Right lung

Left lung

Heart

Diaphragm

The heart is simply an efficient and powerful pump, composed almost entirely of muscle, that beats tirelessly and continuously to keep blood circulating throughout the body. About the size of a clenched fist and shaped like a blunt cone, the heart is situated within the mediastinum in the thoracic cavity, flanked by the lungs and resting on the central tendon of the diaphragm. About two-thirds of the heart lies to the left of the midline of the chest, with the remaining third lying to the right. The organ has four hollow chambers, two atria that "receive" blood into the heart, and two ventricles, which pump approximately 1,000 gal (3,800 L) of blood back out into the vascular system every 24 hours.

Body system:	cardiovascular system
Location:	in the mediastinum between the lungs and behind the sternum
Function:	pumps deoxygenated blood to the lungs for oxygenation; pumps blood rich in oxygen and nutrients to the body's tissues
Components:	pericardium, two atria, two ventricles, valves
Related parts:	arteries and veins, lungs

The Pericardium

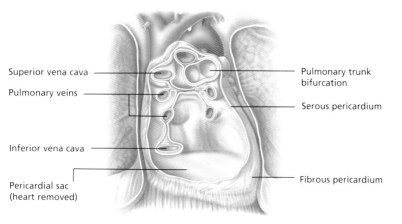

Superior vena cava

Pulmonary veins

Inferior vena cava

Pericardial sac
(heart removed)

Pulmonary trunk
bifurcation

Serous pericardium

Fibrous pericardium

The heart is enclosed within a triple-walled bag of connective tissue called the pericardium. The pericardium has two parts—the fibrous pericardium, which is the tough outer layer, and the serous pericardium, a thin membrane that covers and surrounds the heart itself. The fibrous pericardium is strong enough to provide some protection from trauma and, because it is not elastic, prevents the heart from expanding with blood beyond a safe limit. The serous pericardium has two layers that are continuous with each other (parietal and visceral layers), between which there is a slitlike cavity filled with a small amount of fluid. This allows the chambers of the heart to move freely within the pericardium as the heart beats.

Body system:	cardiovascular system
Location:	surrounding the heart
Function:	protection, helps to anchor heart to surrounding structures, provides lubrication to enable heart to move within the sac
Components:	fibrous pericardium, serous pericardium
Related parts:	heart, nearby blood vessels, sternum

The Heart Wall

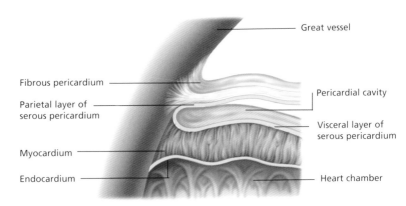

Great vessel

Fibrous pericardium

Parietal layer of serous pericardium

Myocardium

Endocardium

Pericardial cavity

Visceral layer of serous pericardium

Heart chamber

Inside the pericardial cavity, the heart wall is made up of three layers: the epicardium, myocardium, and endocardium. The epicardium is the visceral layer of the serous pericardium that covers the outer surface of the heart and is attached firmly to it. The central layer, the myocardium, makes up the bulk of the heart wall and is composed of specialized cardiac muscle fibers that are present only in the heart and are adapted for the special role they play. The muscle fibers in the myocardium are supported and held together by interlocking fibers of connective tissue. The endocardium is a delicate membrane, formed by a very thin layer of cells, and lines the inner surface of the heart chambers and valves.

Body system:	cardiovascular system
Location:	outer wall surrounding the heart
Function:	forms bulk of the heart; muscle fibers contract to squeeze blood into the circulation
Components:	epicardium, myocardium, endocardium
Related parts:	pericardium, heart valves

Chambers of the Heart

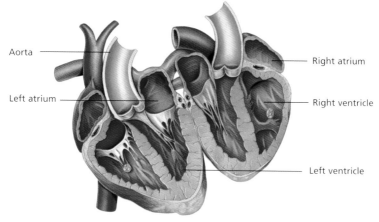

Aorta

Left atrium

Right atrium

Right ventricle

Left ventricle

The heart is divided into left and right sides, each of which has two chambers, or cavities (shown here cut away). The upper chambers are known as atria, and these lead into the lower ventricles through valves that prevent the backflow of blood. The two ventricles make up the bulk of the muscle of the heart. The left ventricle is larger and more powerful than the right and receives oxygenated blood from the lungs via the left atrium. Powerful contractions then pump the blood up out of the heart and into the aorta, the main artery of the body, to be distributed through the arterial system. The right ventricle receives "used" blood from the right atrium and then pumps it out toward the lungs for oxygenation.

Body system:	cardiovascular system
Location:	in the heart
Function:	receive circulating blood and channel it out, either to the lungs for reoxygenation or to the body's tissues
Components:	left and right atria, left and right ventricles
Related parts:	veins and arteries

Ventricular Walls

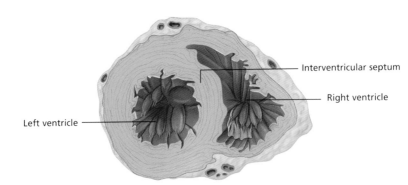

Interventricular septum

Right ventricle

Left ventricle

The walls of the left ventricle are twice as thick as those of the right, and they form a rough circle in cross-section, whereas the right ventricle appears compressed by the more muscular left chamber. The difference in muscle thickness between the two chambers reflects the pressure required to empty each when the muscle contracts. Arising from the walls of both ventricles are papillary muscles, which taper to a point and bear tendinous chords (chordae tendinae) that attach to the tricuspid and mitral valves to stabilize them. The inner surfaces of the ventricles, especially where blood enters, are roughened by irregular ridges of muscle, the trabeculae carnae, which give way to smoother surfaces near the outflow tracts.

Body system:	cardiovascular system
Location:	in the heart, below the atria
Function:	muscular walls pump deoxygenated blood to the lungs or oxygenated blood throughout the body
Components:	muscle, chordae tendinae, valves
Related parts:	atria, arteries, and veins, lungs

The Atria

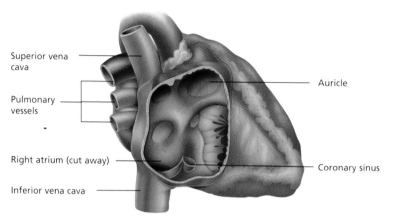

Superior vena cava

Pulmonary vessels

Right atrium (cut away)

Inferior vena cava

Auricle

Coronary sinus

The atria are the two smaller, thin-walled chambers of the heart. They sit above the ventricles, separated from them by the atrioventricular valves. All the venous blood from the body is delivered to the right atria via the two great veins, the superior and inferior venae cavae. The coronary sinus also drains venous blood from the heart muscle into this chamber. The left atria is smaller than the right and receives oxygenated blood from the lungs via the pulmonary veins. The interior of each atrium is smooth, except for the anterior walls, which are ridged by muscle bundles called pectinates. Inside each atria is an appendage called an auricle (so called because it resembles an ear), which increases the surface area.

Body system:	cardiovascular system
Location:	in the heart, above the ventricles
Function:	receive venous blood from the body and the muscle of the heart itself, and oxygenated blood from the lungs
Components:	muscle tissue, valves
Related parts:	ventricles, arteries, and veins, lungs

Valves of the Heart

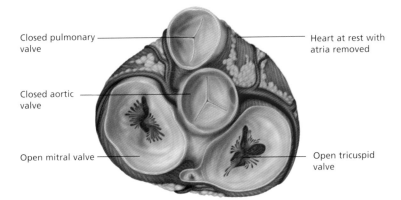

Closed pulmonary valve

Closed aortic valve

Open mitral valve

Heart at rest with atria removed

Open tricuspid valve

Blood flows through the heart in one direction only, and backflow is prevented by the four heart valves. On the right side of the heart, the tricuspid valve lies between the atrium and the ventricle, and the pulmonary valve is located at the junction of the ventricle and the pulmonary trunk. On the left side, the mitral valve separates the atrium and ventricle, while the aortic valve is situated between the ventricle and the aorta. The tricuspid and mitral valves, also known as the atrioventricular valves, are composed of tough connective tissue covered with endocardium, the thin layer of cells that lines the whole heart. The tricuspid valve has three cusps (flaps) that open and shut, while the mitral valve has only two.

Body system :	cardiovascular system
Location:	between the atria and ventricles of the heart, and between the ventricles and the great vessels
Function:	ensure that blood flows in one direction only through the heart
Components:	tricuspid, pulmonary, aortic, and mitral valves
Related parts:	atria, ventricles, chordae tendinae

The Semilunar Valves

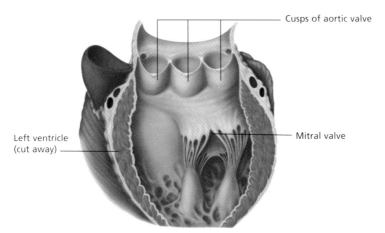

Cusps of aortic valve

Left ventricle
(cut away)

Mitral valve

Also known as the semilunar valves, the aortic and pulmonary valves guard the routes of exit of blood from the heart (the aorta and the pulmonary trunk respectively), preventing backflow of blood as the ventricles relax after a contraction. Each of these two valves is composed of three pocketlike semilunar cusps, or flaps, (so named because they resemble half moons), which have a core of connective tissue covered by a lining of endothelium and are attached to the artery wall. This smooth lining ensures an ideal surface for the passage of blood. The aortic valve is much stronger and more robust than the pulmonary valve, because it has to cope with the higher pressures of the arterial circulation.

Body system:	cardiovascular system
Location:	at the junction of the ventricles and the aorta and pulmonary trunk
Function:	prevent backflow of blood into the heart
Components:	three semilunar cusps
Related parts:	ventricles, aorta, pulmonary trunk

Action of the Valves

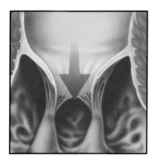
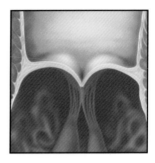

Open valve   **Closed valve**

When the atria contract, blood passes through the open tricuspid and mitral valves into the ventricles. As the ventricles contract in turn, the sudden rising pressure of blood within each ventricle causes the valves to close, so preventing backflow of blood into the atria. The pull of the chordae tendinae (tendinous cords that anchor to muscles protruding from the ventricle walls) steadies the valves and enables them to withstand the pressure of the blood in the ventricle. When the atrioventricular valves are closed, the blood must travel up and out through the semilunar valves into the pulmonary trunk and the aorta. These valves are forced open by the high pressure flow of blood but snap shut again as the ventricles relax.

Body system:	cardiovascular system
Location:	in the heart between the atria and ventricles, and the ventricles and aorta and pulmonary trunk
Function:	prevent the backflow of blood
Components:	tricuspid, mitral, pulmonary, and aortic valves
Related parts:	heart, aorta, pulmonary trunk

The Great Vessels

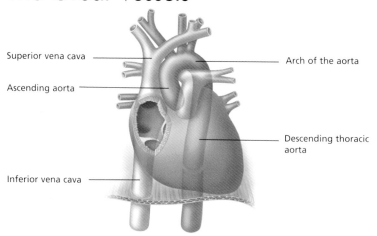

Superior vena cava

Ascending aorta

Inferior vena cava

Arch of the aorta

Descending thoracic aorta

Blood is delivered to the heart from the body tissues by two large veins—the superior and inferior venae cavae—and pumped out into the aorta; these vessels are collectively known as the great vessels. The superior vena cava drains blood from the upper body to the right atrium of the heart and is formed by the union of the right and left brachiocephalic veins. The inferior vena cava is the widest vein in the body, but only its last section lies within the thorax as it passes up through the diaphragm to deliver blood to the right atrium. The aorta is the largest artery in the body, with a diameter of 1 in (2.5 cm) in adults. Its thick wall contains elastic tissue that allows it to expand as blood is pumped into it at high pressure.

Body system:	cardiovascular system
Location:	superior and inferior vena cava drains into right atrium, aorta leaves left ventricle
Function:	superior and inferior venae cavae deliver deoxygenated blood from the body's tissues back to the heart; aorta carries oxygenated blood from the heart to the body
Components:	superior and inferior venae cavae, aorta
Related parts:	heart, other veins and arteries

Coronary Arteries

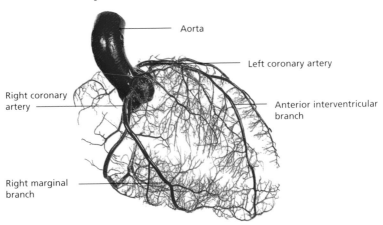

Aorta

Left coronary artery

Right coronary artery

Anterior interventricular branch

Right marginal branch

Although the heart is continually filled with blood, its muscle and pericardium need their own blood supply to provide oxygen and nutrients to the tissues. This is provided by the two coronary arteries, right and left, which arise from the ascending aorta just above the aortic valve and run around the heart beneath the epicardium embedded in fat. The left coronary artery supplies the left side of the heart and branches into two main vessels, the anterior interventricular artery and the circumflex artery. The right coronary artery also divides into two branches, the marginal and posterior interventricular arteries. As these larger arteries encircle the heart, they give off numerous smaller branches to provide a rich blood supply.

Body system:	cardiovascular system
Location:	originate in the ascending aorta and encircle the heart
Function:	supply the heart muscle and tissues with oxygen and nutrients
Components:	left and right coronary arteries
Related parts:	aorta, heart muscle, and pericardium

Conduction System

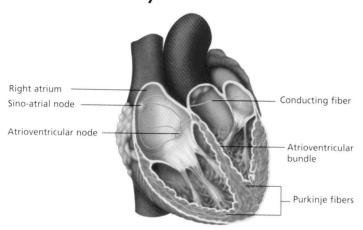

Right atrium

Sino-atrial node

Atrioventricular node

Conducting fiber

Atrioventricular bundle

Purkinje fibers

Normally, the heart contracts at a rate of about 75 beats a minute. Within its muscular walls, an independent conducting system sets the pace and ensures that the chambers contract in a coordinated way, pushing blood through the heart and out into the great vessels. The sino-atrial node—a collection of cells in the wall of the right atrium—is the heart's natural pacemaker. These cells generate electrical impulses that pass to the atria and then to the atrioventricular (AV) node in the floor of the right atrium. Impulses from the AV node are passed down the central septum via the atrioventricular bundle, which splits into two branches. These continue downward, becoming the Purkinje fibers present in the ventricular walls.

Body system:	cardiovascular system
Location:	wall and floor of right atrium, septum, ventricular walls
Function:	generate and transmit regular electrical impulses to stimulate the heart's muscle cells to contract
Components:	sino-atrial node, atrioventricular node, atrioventricular bundle, right and left bundle branches, Purkinje fibers
Related parts:	heart muscle cells

The Cardiac Cycle

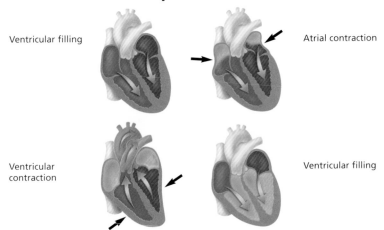

Ventricular filling

Atrial contraction

Ventricular
contraction

Ventricular filling

The cardiac cycle is the series of pressure and volume changes within the heart that causes blood to be pumped around the body. It occurs in two phases: a period when the heart muscle contracts (systole) and a period when it is relaxed (diastole). During diastole, the tricuspid and mitral valves are open. Blood from the circulation fills the atria and passes through the open valves to the ventricles, causing the atrioventricular valves to close. As systole begins, the sino-atrial node stimulates the atria, which contract, forcing more blood through into the ventricles. As the atria relax and the wave of impulses reaches the ventricles, they too contract and force blood out through the semilunar valves and into the circulation.

Body system:	cardiovascular system
Location:	the heart
Function:	to enable blood to be pumped through the circulation
Components:	systole (contraction) and diastole (relaxation)
Related parts:	atria and ventricles, valves, great vessels

The Shoulder Joint

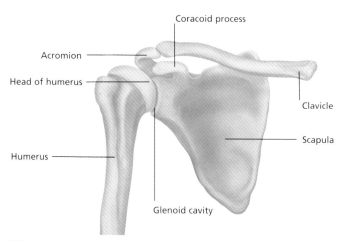

Coracoid process

Acromion

Head of humerus

Clavicle

Scapula

Humerus

Glenoid cavity

The glenohumeral, or shoulder, joint is the point of articulation between the glenoid cavity of the scapula (shoulder blade) and the head of the humerus, the bone of the upper arm. It is a ball-and-socket synovial joint structured to allow the arm a wide range of movement. The glenoid cavity of the scapula provides only a shallow socket and, therefore, strong muscles and tendons are necessary to hold the bones firmly together. The shoulder joint is surrounded by a loose capsule of fibrous tissue lined by a synovial membrane. This membrane secretes synovial fluid, a viscous liquid that lubricates and nourishes the joint. A thin layer of smooth articular cartilage allows the bones to slip over each other with minimum friction.

Body system:	musculoskeletal system
Location:	at the junction of the scapula and the head of the humerus
Function:	to enable a wide range of movement in the upper limbs
Components:	glenoid cavity of scapula, head of humerus, articular cartilage, synovial membrane
Related parts:	arms, muscles, scapula

Bursae of the Shoulder Joint

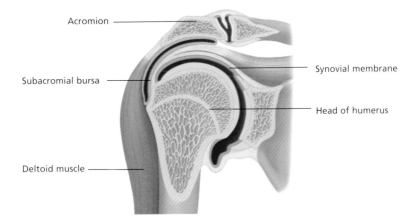

Acromion

Synovial membrane

Subacromial bursa

Head of humerus

Deltoid muscle

A bursa is a flattened fibrous sac, lined with synovial membrane, which contains a small amount of viscous synovial fluid. Bursae reduce the friction between structures that necessarily come into contact with each other during normal movement. They are located at various points around the body where ligaments, muscle, and tendons rub against the bone. The shoulder joint has several important bursae. The subcapsular bursa protects the tendon of the subscapularis muscle as it passes over the neck of the scapula. The subacromial bursa lies above the glenohumeral joint beneath the acromion and the coraco-acromial ligament. This bursa allows free movement of the muscles that pass underneath it.

Body system :	musculoskeletal system
Location:	at points in the shoulder joint where muscles and tendons rub against bone
Function:	reduce friction between two adjacent structures
Components:	fibrous sac containing synovial fluid
Related parts:	bones, muscles, tendons, ligaments

Ligaments of the Shoulder

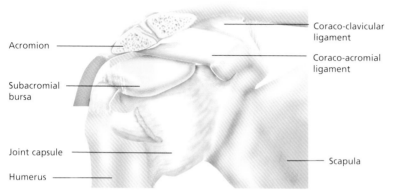

Acromion

Subacromial bursa

Joint capsule

Humerus

Coraco-clavicular ligament

Coraco-acromial ligament

Scapula

The ligaments around any joint contribute to its stability by holding the bones firmly together. Stability is particularly important in the shoulder joint, which is unusually shallow to allow a wide range of movement. The main stabilizers are the surrounding muscles but ligaments also play a role. The fibrous joint capsule has ligaments within it that help to strengthen the joint, including the glenohumeral ligaments (three weak fibrous bands that reinforce the front of the capsule), while the shoulder joint is strengthened by various ligaments such as the coraco-acromial ligament. The transverse humeral ligament runs from the greater to the lesser tuberosity of the humerus, creating a tunnel for the biceps brachii tendon.

Body system:	musculoskeletal system
Location:	surrounding and within the shoulder joint
Function:	to stabilize the joint and hold the head of the humerus in the glenoid cavity
Components:	glenohumeral, coraco-acromial, and transverse humeral ligaments
Related parts:	bones of shoulder joint

Muscles of the Shoulder

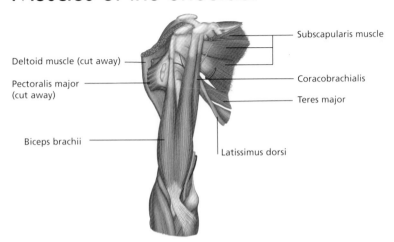

Subscapularis muscle

Deltoid muscle (cut away)

Pectoralis major
(cut away)

Coracobrachialis

Teres major

Biceps brachii

Latissimus dorsi

The shoulder is a ball-and-socket joint that allows 360° of movement to give maximum flexibility. Shoulder movements take place around three axes, allowing flexion and extension, abduction and adduction, and medial and lateral rotation. Many of the muscles involved in these movements are attached to the pectoral girdle (clavicles and scapulae). For example, the powerful deltoid muscle, which enables a number of movements, is attached to the acromion process, which projects from the scapula over the shoulder joint. Some muscles arise directly from the trunk (pectoralis major and latissimus dorsi), while others influence the movement of the humerus even though they are not attached to it (such as the trapezius).

Body system:	musculoskeletal system
Location:	surrounding the shoulder joint
Function:	to enable maximum flexibility in the shoulder joint and to provide stability
Components:	deltoid, pectoralis major, biceps brachii, subscapularis, coracobrachialis, teres major
Related parts:	pectoral girdle, humerus, elbow

Rotation of the Arm

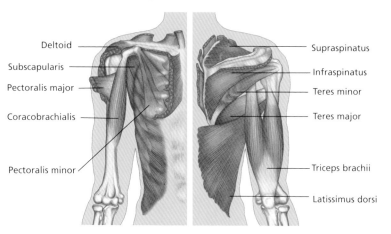

Deltoid

Subscapularis

Pectoralis major

Coracobrachialis

Pectoralis minor

Supraspinatus

Infraspinatus

Teres minor

Teres major

Triceps brachii

Latissimus dorsi

The pectoralis major, the anterior fibers of the deltoid, the teres major, and the latissimus dorsi muscles all enable medial rotation of the humerus. The most powerful medial rotator, however, is the subscapularis. This muscle occupies the entire front surface of the scapula, and it attaches to the joint capsule around the lesser tuberosity of the humerus. Subscapularis is one of a set of four short muscles, collectively known as the "rotator cuff," which attach to and strengthen the joint capsule. In addition, they pull the humerus into the socket of the joint increasing contact of the bones and contributing to the stability of the shoulder. The other muscles of the rotator cuff are the supraspinatus, infraspinatus, and the teres minor.

Body system:	musculoskeletal system
Location:	surrounding the shoulder joint and in the back
Function:	to medially rotate (turn on an axis) the arm
Components:	pectoralis major, deltoid, teres major, latissimus dorsi, subscapularis, supraspinatus, infraspinatus, teres minor
Related parts:	other muscles of back and arms, pectoral girdle, humerus

Axilla

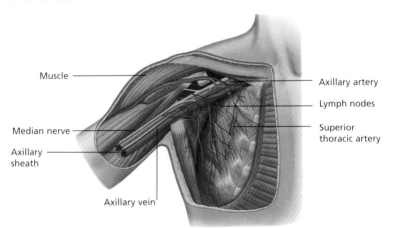

Muscle

Axillary artery

Lymph nodes

Median nerve

Superior thoracic artery

Axillary sheath

Axillary vein

A dense network of blood vessels, nerves, and lymphatic vessels serving the upper limb passes through the axilla (armpit), a roughly pyramidal space located at the junction of the upper arm and thorax. The axillary artery and its branches supply oxygenated blood to the upper limb and the axillary vein runs parallel to the artery. The nerves that lie in the axilla are all part of a complex network known as the brachial plexus. The blood vessels and nerves are protected and enclosed by a tube of strong connective tissue, the "axillary sheath," which is important as the axilla is vulnerable to trauma from below. Within the fatty connective tissue of the axilla lie groups of lymph nodes connected by lymphatic vessels.

Body system:	various systems
Location:	at the junction of the upper limbs and thorax
Function:	site of intersection for important blood vessels, nerves, and lymphatic vessels
Components:	fatty and connective tissue, axillary artery and vein, brachial plexus, lymphatic vessels and nodes
Related parts:	other structures in the upper limbs and thorax

The Clavipectoral Fascia

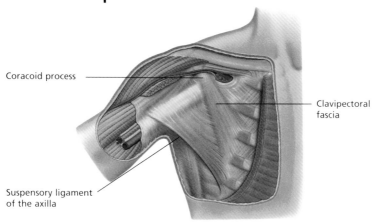

Coracoid process

Clavipectoral fascia

Suspensory ligament of the axilla

The clavipectoral fascia is a sheet of strong connective tissue attached at its upper border to the coracoid process of the scapula and clavicle. The fascia descends to enclose the subclavius and pectoralis minor muscles and then joins with the overlying axillary fascia in the base of the axilla (armpit). The part of the clavipectoral fascia that lies above the pectoralis minor is known as the costacoracoid membrane and is pierced by the nerve that supplies the overlying pectoralis minor muscle. Below the pectoralis minor muscle, the fascia becomes the suspensory ligament of the axilla, which attaches to the skin of the armpit and is responsible for pulling that skin up when the arm is raised.

Body system:	musculoskeletal system
Location:	extending from the upper chest across the axilla
Function:	protects contents of the axilla, acts as a suspensory ligament
Components:	connective tissue
Related parts:	pectoral girdle, suspensory ligament

The Humerus

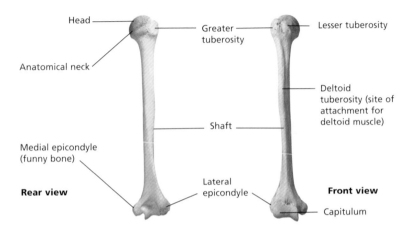

Head

Greater tuberosity

Lesser tuberosity

Anatomical neck

Deltoid tuberosity (site of attachment for deltoid muscle)

Shaft

Medial epicondyle (funny bone)

Rear view

Lateral epicondyle

Front view

Capitulum

The humerus, a typical "long bone," is the bone of the upper arm. At the top of the humerus is the smooth head (epiphysis) that fits into the glenoid cavity of the scapula at the shoulder joint. Behind the head is a shallow indentation known as the anatomical neck, which separates the head from two bony prominences, the greater and lesser tuberosities (sites for muscle attachment). The shaft is the long smooth length of bone that runs the length of the upper arm. Ridges at each side of the lower shaft terminate in the prominent medial (inner) and lateral epicondyles. There are two main parts to the articular surface: the trochlea, which articulates with the ulna, and the capitulum, which articulates with the radius.

Body system:	musculoskeletal system
Location:	extends from the shoulder to the elbow
Function:	provides stable frame for the arm, sites for muscle attachment, contains bone marrow, which produces blood cells
Components:	epiphyses, greater and lesser tuberosities, shaft, medial and lateral epicondyles, capitulum, trochlea
Related parts:	pectoral girdle, ulna, radius, muscles of upper arm

The Radius and the Ulna

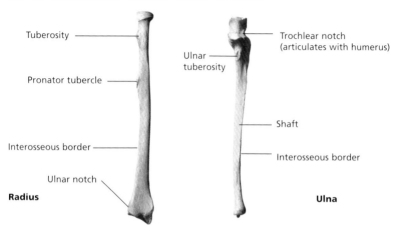

Tuberosity

Pronator tubercle

Interosseous border

Ulnar notch

Radius

Trochlear notch
(articulates with humerus)

Ulnar
tuberosity

Shaft

Interosseous border

Ulna

Two parallel long bones, the radius and the ulna, run the length of the forearm, connecting with the humerus at their upper ends and forming joints in the wrist at the lower ends. The ulna lies on the same side as the little finger, while the radius is on the same side as the thumb. The radio-ulnar joints allow the ulna and radius to rotate around each other in the movements peculiar to the forearm known as pronation (rotating the forearm so the palm faces down) and supination (palm faces up). The ulna is longer than the radius and is the main stabilizing bone of the forearm. The ulna is the forearm bone that contributes most to the elbow joint, while the radius plays a major part in forming the wrist joints.

Body system:	musculoskeletal system
Location:	between the elbow and wrist joints
Function:	provide stable frame for the arm, sites for muscle attachment, allow flexibility of movement
Components:	head, neck, shaft, tuberosities, notches, interosseous border
Related parts:	interosseous membrane, humerus, wrist joints, muscles

Interosseous Membrane

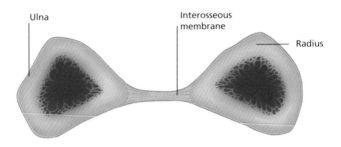

Cross-section through radius and ulna

The radius and ulna are connected along their length by a thin sheet of tough, fibrous connective tissue called the interosseous membrane. The membrane is broad enough to enable a certain amount of movement between the bones during the actions of supination and pronation (turning the palm up and down) and is strong enough to provide sites for the attachment of some of the deep muscles of the forearm. The interosseous membrane has an important role to play in the transmission of force through the forearm. If force were to be applied to the wrist (as in a fall) the fibers of the membrane lie in such a way that the impact is directed up to the elbow and upper arm, rather than absorbed by the wrist.

Body system:	musculoskeletal system
Location:	between the radius and ulna
Function:	stabilizes the bones, securing them together; allows flexibility of movement, protects against a forceful blow
Components:	tough fibers (connective tissue)
Related parts:	radius, ulna, bones of wrist

The Elbow

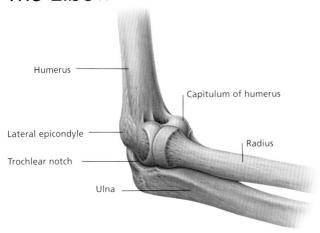

Humerus

Capitulum of humerus

Lateral epicondyle

Radius

Trochlear notch

Ulna

The elbow is a synovial (fluid-filled) joint between the lower end of the humerus and the upper ends of the ulna and radius. It is a hinge joint and, therefore, moves in one plane only (bending and straightening the arm) so that the structure is very stable. The main stability of the elbow comes from the size and depth of the trochlear notch of the ulna, which effectively grips the lower end of the humerus like a wrench. All the opposing joint surfaces are covered with smooth articular cartilage to reduce friction between the bones during movement, and the entire joint is surrounded by a fibrous capsule, which extends down from the articular surfaces of the humerus to the upper end of the ulna.

Body system:	musculoskeletal system
Location:	between the lower end of the humerus and the upper ends of the radius and ulna
Function:	enables flexion (bending) and extension (straightening) of the arm
Components:	lower end of humerus, head of radius and ulna, fibrous capsule
Related parts:	ligaments and muscles

Ligaments of the Elbow

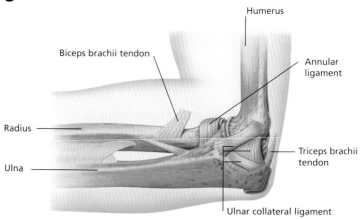

Humerus

Biceps brachii tendon

Annular ligament

Radius

Triceps brachii tendon

Ulna

Ulnar collateral ligament

The elbow is strengthened and supported at each side by strong collateral ligaments, which are thickenings of the joint capsule. The radial collateral ligament (not shown) is a fan-shaped structure that originates from the lateral epicondyle—a bony prominence on the outer side of the lower end of the humerus (upper arm bone). This ligament runs downward to blend with the annular ligament, which encircles the head of the radius. The ulnar collateral ligament runs between the medial (inner) epicondyle of the humerus and the upper end of the ulna (lower arm bone) and is made up of three parts that together form a triangle. The tendons of the triceps and biceps brachii also help stabilize the elbow.

Body system:	musculoskeletal system
Location:	surrounding the elbow joint
Function:	strengthen and support the elbow joint
Components:	radial collateral, ulnar collateral, and annular ligaments
Related parts:	ulna, radius, humerus

Anterior Muscles of the Upper Arm

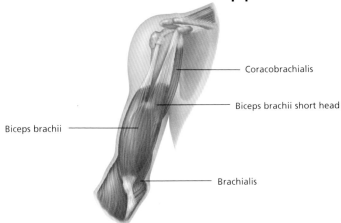

Coracobrachialis

Biceps brachii short head

Biceps brachii

Brachialis

The muscles of the upper arm can be divided into two distinct areas—the anterior and posterior compartments. The muscles of the anterior compartment are all flexors, that is, they bend the arm. The muscle that gives the upper arm its distinctive bulge is the biceps brachii, which arises from two heads that join together to form the body of the muscle. As well as acting to flex the forearm, the biceps is also a supinator and helps to turn the forearm so that the palm is facing upward. The brachialis muscle is flatter than the biceps, lying directly beneath it, and is the main flexor muscle of the elbow. The coracobrachialis helps flex the upper arm at the shoulder and to pull it back in line with the body.

Body system:	musculoskeletal system
Location:	in the anterior compartment of the upper arm
Function:	flex the arm
Components:	biceps brachii, brachialis, coracobrachialis
Related parts:	other arm muscles, humerus, elbow, pectoral girdle

Posterior Muscles of the Upper Arm

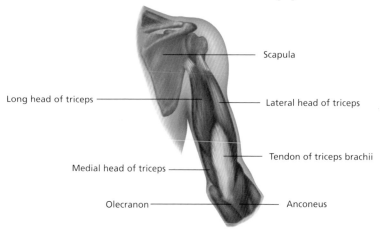

Scapula

Long head of triceps

Lateral head of triceps

Tendon of triceps brachii

Medial head of triceps

Olecranon

Anconeus

The posterior compartment has only two muscles—one major muscle, the triceps brachii, and the small anconeus muscle. The triceps is a large bulky muscle that lies behind the humerus (bone of the upper arm) and, as its name implies, has three heads: the long head, the lateral head, and the medial head. The three heads converge in the middle of the upper arm on a wide flattened tendon, which passes down, over a small bursa, to attach to the olecranon process of the ulna (bone in the lower arm). The main action of the triceps is to extend the elbow, thereby straightening the arm. In addition, the long head of the triceps and the small anconeus muscle behind the elbow joint both help to stabilize the joint.

Body system :	musculoskeletal system
Location:	the posterior compartment of the upper arm
Function:	extend the elbow, straightening the arm; stabilize elbow joint
Components:	triceps brachii, anconeus
Related parts:	elbow joint, humerus, scapula

Flexor Muscles of the Forearm

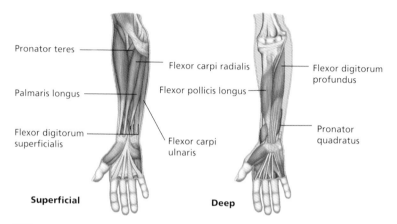

Pronator teres

Flexor carpi radialis

Palmaris longus

Flexor pollicis longus

Flexor digitorum superficialis

Flexor carpi ulnaris

Flexor digitorum profundus

Pronator quadratus

Superficial

Deep

The superficial and deep flexor muscles of the front compartment of the forearm act to flex the hand, wrist, and fingers. The superficial group contains five muscles, all of which originate at the medial epicondyle of the humerus, where their fibers merge to form the common flexor tendon. These help to pronate the forearm, flex the elbow, and act to produce flexion, abduction, and adduction of the wrist. One of the small superficial muscles, the palmaris longus, is absent in 14 percent of people. The deep layer of flexor compartment consists of three muscles: the flexor digitorum profundus, the flexor pollicis longis, and the pronator quadratus. These lie close to the bones and act to move the fingers and thumbs and to pronate the forearm.

Body system:	musculoskeletal system
Location:	between the lower end of the humerus and the bones of the wrist
Function:	enables flexion (bending) of the arm and digits
Components:	pronator teres, flexor carpi radialis, palmaris longus, flexor carpi ulnaris, flexor digitorum superficialis, flexor digitorum profundus, flexor pollicis longus, pronator quadratus
Related parts:	humerus, bones of forearm, wrist, and hand, tendons and ligaments

Extensor Muscles of the Forearm

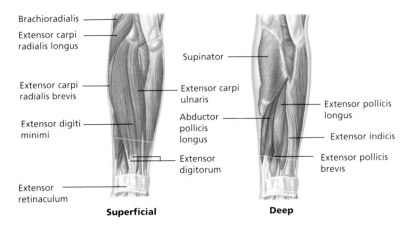

Brachioradialis

Extensor carpi radialis longus

Supinator

Extensor carpi radialis brevis

Extensor carpi ulnaris

Extensor pollicis longus

Abductor pollicis longus

Extensor indicis

Extensor digiti minimi

Extensor digitorum

Extensor pollicis brevis

Extensor retinaculum

Superficial

Deep

Working together with the flexor muscles, the extensor muscles of the forearm enable a wide range of movements in the wrist, hand, fingers, and thumb. The posterior (rear) extensor compartment is separated from the flexor muscles by the radius and ulna and by sheets of connective tissue. The extensor muscles can be divided into three groups according to their functions: muscles that move the hand or the wrist (extensor carpi muscles); muscles that straighten the fingers (extensor digiti minimi, extensor digitorum); and muscles that act only on the thumb (brachioradialis). Lying closer to the underlying bones, the deep extensor layer includes muscles that act upon the thumb and the little finger individually.

Body system:	musculoskeletal system
Location:	between the elbow and the wrist
Function:	enable a range of movement in the forearm, hand, and digits
Components:	extensor carpi radialis longus, extensor carpi radialis brevis, extensor carpi ulnaris, extensor digitorum, extensor digiti minimi, extensor indicis, brachioradialis, abductor pollicis longus, etc.
Related parts:	elbow, ulna, radius, wrist joints, bones in hand and digits

Attachments of Extensor Tendons

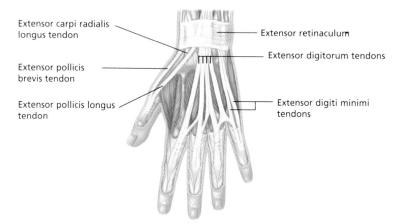

Extensor carpi radialis longus tendon

Extensor retinaculum

Extensor digitorum tendons

Extensor pollicis brevis tendon

Extensor pollicis longus tendon

Extensor digiti minimi tendons

Most of the muscles of the posterior extensor compartment of the forearm terminate in long tendons that pass down over the back of the wrist to attach to the bones of the hands and fingers. In this way, muscles in the forearm can bring about extension (straightening and bending back) of the hand and fingers by "remote control," thereby allowing the hand itself to be relatively muscle-free and therefore less bulky. The site of attachment of each tendon determines which joint of the hand will be straightened when that muscle contracts. As the extensor tendons pass over the back of the wrist, they pass under the extensor retinaculum, a band of connective tissue that holds them against the joint as the hand moves.

Body system:	musculoskeletal system
Location:	back of the wrist and hand
Function:	attach muscles to bone, bring about movement in wrist, hand, digits
Components:	extensor carpi radialis longus tendon, extensor pollicis brevis tendon, extensor pollicis longus tendon, extensor retinaculum, extensor digitorum tendons, extensor digiti minimi tendons
Related parts:	fingers, thumbs, wrist joints

Arteries of the Arm

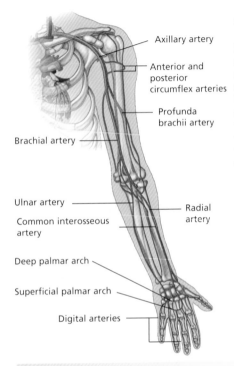

Axillary artery

Anterior and posterior circumflex arteries

Profunda brachii artery

Brachial artery

Ulnar artery

Common interosseous artery

Radial artery

Deep palmar arch

Superficial palmar arch

Digital arteries

The main blood supply to the arm is provided by the brachial artery, a continuation of the axillary artery, which runs down the inner side of the upper arm. From it arises many smaller branches that supply muscles and the humerus (bone in the upper arm). The largest of these is the profunda brachii artery, which supplies the muscles that straightens the elbow. The brachial artery divides below the elbow into the radial and ulnar arteries. Both these arteries run the length of the forearm along the radius and ulna (the two bones in the forearm) and "pulses" can be felt over them at the inner wrist. The hand receives a rich blood supply from the deep and superficial palmar arteries, the end branches of the radial and ulnar arteries.

Body system:	cardiovascular system
Location:	travel the length of the arm from the axillary artery
Function:	to provide blood rich in oxygen and nutrients to the tissues and structures of the arm
Components:	anterior and posterior circumflex, profunda brachii, radial, digital, deep and superficial palmar arch, ulnar, common interosseous, brachial, axillary arteries
Related parts:	humerus, ulna, radius, subclavian artery

Veins of the Arm

Venous drainage of the upper limb is achieved by two interconnecting series of veins, the deep and superficial systems. Deep veins run alongside the arteries, while superficial veins lie in the subcutaneous tissue. In most cases, the deep veins are paired veins that lie on either side of the artery they accompany, making frequent connections and forming a network around the arteries. The radial and ulnar veins arise from the palmar venous arches of the hand and run up the forearm to merge at the elbow, forming the brachial vein. The two main superficial veins, the cephalic and basilic veins, originate at the dorsal venous arch of the hand. The cephalic vein runs under the skin along the radial side of the forearm and the basilic vein runs up the ulnar side, crossing the elbow to lie along the border of the biceps muscle.

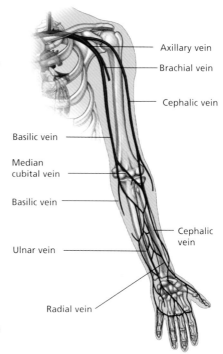

Axillary vein

Brachial vein

Cephalic vein

Basilic vein

Median cubital vein

Basilic vein

Cephalic vein

Ulnar vein

Radial vein

Body system:	cardiovascular system
Location:	travel the length of the arm from the axillary vein
Function:	to drain deoxygenated blood from the tissues and structures of the arm and return it to the heart
Components:	radial, ulnar, brachial, basilic, cephalic, median cubital, axillary veins
Related parts:	humerus, ulna, radius, subclavian vein

The Brachial Plexus

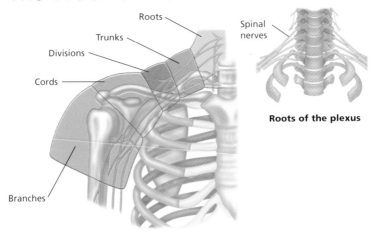

Roots

Trunks

Divisions

Cords

Branches

Spinal nerves

Roots of the plexus

The brachial plexus is a complex network of nerves originating in the spinal cord, from which arises most of the nerves supplying the upper limbs. The plexus is formed by the intermixing of the ventral rami (lower branch of each spinal nerve) of four cervical nerves and most of the first thoracic ramus. It is divided into sections which, starting from the spine, are known as the roots, trunks, divisions, cords, and branches. The roots lie within the neck to either side of the spinal column and the three trunks are located above the clavicle. The divisions arise from the trunks and pass behind the clavicle, entering the axilla and from there become cords. The terminal branches of the brachial plexus pass into the upper limb.

Body system:	nervous system
Location:	extends from the spinal cord in the neck to the upper limb
Function:	from it arises most of the nerves that supply the upper limb
Components:	roots, three trunks, divisions, three cords, terminal branches
Related parts:	spinal cord, brain, nerves of upper limb

Nerves of the Arm

The nerve supply to the arm is provided by four main nerves and their branches. These receive sensory information from the hand and arm, and they also innervate the numerous muscles of the upper limb. The radial and musculo-cutaneous nerves supply muscles and skin of all parts of the arm, while the median and ulnar nerves only supply structures below the elbow. The radial nerve is of great importance because it is the main supplier of innervation to the extensor muscles, which straighten the elbow, wrist, and fingers. It is the largest branch of the brachial plexus, a network of nerves in the spinal cord in the neck, from which most of the nerves that supply the arm arise. Near the lateral epicondyle in the elbow, the radial nerve divides into its two terminal branches, the superficial and deep terminal branches.

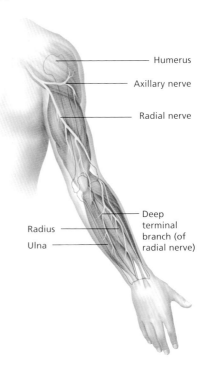

- Humerus
- Axillary nerve
- Radial nerve
- Deep terminal branch (of radial nerve)
- Radius
- Ulna

Body system:	nervous system
Location:	extend from the brachial plexus to the hand
Function:	provide a sensory and motor nerve supply to the upper limb
Components:	radial, musculocutaneous, median, and ulnar nerves
Related parts:	brachial plexus, axillary nerve, muscles and tissues of the upper limb

Median and Ulnar Nerves

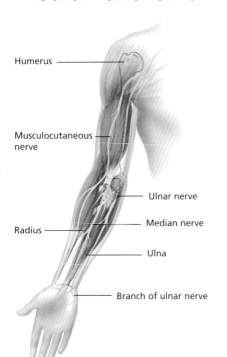

Humerus

Musculocutaneous nerve

Radius

Ulnar nerve

Median nerve

Ulna

Branch of ulnar nerve

The median nerve is the main nerve of the front of forearm. It arises from the brachial plexus and runs downward centrally to the elbow. At the wrist, the median nerve passes through the carpal tunnel and ends in branches that supply some of the small muscles of the hand, as well as the skin over the thumb and some neighboring fingers. The ulnar nerve passes down along the humerus to the elbow and then loops behind the medial epicondyle just beneath the skin where it can easily be felt. At this point, the nerve is vulnerable to injury and when knocked, can cause tingling in the hand. Before it enters the hand, branches from the ulnar nerve go to the elbow, two of the muscles of the forearm and several areas of skin.

Body system :	nervous system
Location:	travel the length of the arm from the brachial plexus
Function:	provide motor and sensory nerve supply to the structures and tissues of the upper limb
Components:	median and ulnar nerves and branches
Related parts:	brachial plexus, axillary nerve, muscles and tissues of the upper limb

The Wrist

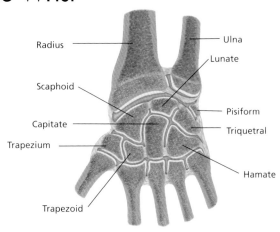

Radius — Ulna

Lunate

Scaphoid

Pisiform

Capitate

Triquetral

Trapezium

Hamate

Trapezoid

The wrist lies between the radius and ulna of the forearm and the bones of the hand. It is made up of eight marble-sized bones, collectively known as the carpals, which are arranged in two rows (the proximal and distal rows). The main joint of the wrist is between the first of these two rows and the lower end of the radius. These bones move together to allow flexibility of the wrist joint and hand. The articular surfaces of the bones are covered in smooth cartilage and enclosed by a synovial membrane, which secretes a viscous fluid to enable the bones to move against each other with minimum friction. The whole joint is covered by a fibrous capsule, which is strengthened by ligaments.

Body system:	musculoskeletal system
Location:	between the radius and ulna and the bones of the hand
Function:	to enable a wide range of movement in the wrist and hand
Components:	triquetral, pisiform, lunate, scaphoid, hamate, capitate, trapezoid, trapezium
Related parts:	ulna, radius, metacarpals

Carpal Tunnel

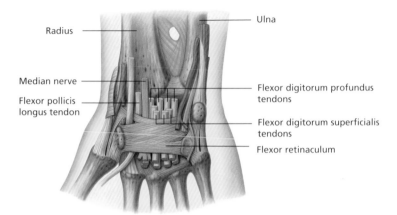

Radius

Ulna

Median nerve

Flexor pollicis longus tendon

Flexor digitorum profundus tendons

Flexor digitorum superficialis tendons

Flexor retinaculum

The eight carpal bones of the wrist fit together to form the shape of an arch. This bony arch is converted into a "tunnel" by a tough band of fibrous tissue, the flexor retinaculum, which lies across the palmar surface and is attached on either side to the bony projections. Through this tunnel, known as the carpal tunnel, run the long tendons of the muscles that flex (bend) the fingers—the flexor digitorum profundus and flexor digitorum superficialis tendons. The flexor retinaculum ensures that these tendons are held close to the wrist even when it is bent, thereby allowing flexion of the fingers in every position. As well as the long flexor tendons, the carpal tunnel also houses the median nerve, the main nerve supplying the hand.

Body system:	musculoskeletal system
Location:	within the wrist
Function:	provides protective and supportive "tunnel" for flexor tendons and the median nerve
Components:	carpal bones, flexor retinaculum
Related parts:	metacarpals, phalanges, muscles of the arm

Ligaments of the Wrist

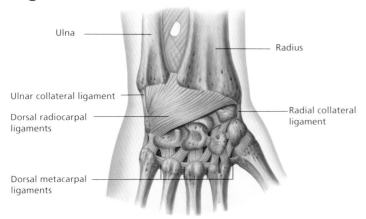

Ulna

Radius

Ulnar collateral ligament

Dorsal radiocarpal ligaments

Radial collateral ligament

Dorsal metacarpal ligaments

The wrist joint itself cannot rotate, and rotation of the hand is achieved by pronation (turning the palm face down) and supination (palm face up) of the forearm. The strong ligaments between the carpal bones and the radius are important—they "carry" the hand around with the forearm during these actions. The palmar radiocarpal ligaments run from the radius to the carpal bones on the palm side of the hand. The fibers are directed so that the hand will go with the forearm during supination. The dorsal radiocarpal ligaments run at the back of the wrist from the radius to the carpal bones and carry the hand back during pronation. Strong collateral ligaments run down each side of the wrist to strengthen the joint.

Body system:	musculoskeletal system
Location:	within the wrist
Function:	enable rotation of the hand; provide stability and support
Components:	palmar radiocarpal, dorsal radiocarpal, dorsal metacarpal, ulnar collateral, radial collateral ligaments
Related parts:	ulna, radius, carpals, metacarpals

Bones of the Hand

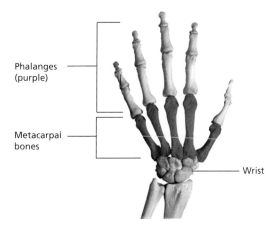

Phalanges (purple)

Metacarpal bones

Wrist

The skeleton of the hand is made up of the eight carpal bones in the wrist, the five metacarpal bones that support the palm and the 14 phalanges, or finger bones. The slender metacarpals radiate out from the wrist toward the fingers to form the support of the palm of the hand. They are numbered from one to five, starting at the thumb. Each of the metacarpals is made up of a shaft and two slightly bulbous ends—in a clenched fist, the knuckles are the heads of the metacarpals. The fingers each contain three phalanges, the first being the largest and the third flattened at the tip to form the skeletal support of the nail bed. The thumb contains only two phalanges but is extremely mobile, allowing a wide range of movement.

Body system:	musculoskeletal system
Location:	in the hand
Function:	provide a framework for the tissues of the hand; enable a wide range of dexterous movements
Components:	metacarpals, phalanges
Related parts:	wrist bones

Finger Joints

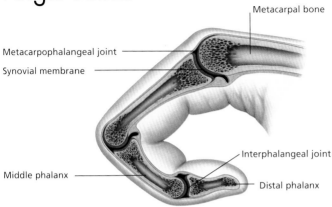

Metacarpal bone

Metacarpophalangeal joint

Synovial membrane

Interphalangeal joint

Middle phalanx

Distal phalanx

The head of each of the five metacarpal bones that support the palm of the hand articulates with the first phalanx of the corresponding finger to form a knuckle. The joints between the metacarpals and the phalanges are "condyloid" synovial joints, a type of joint that allows movement in two planes. They enable the fingers to flex and extend (bend and straighten) and abduct and adduct (move together and apart, spreading the fingers). These movements add to the versatility of the hand, because the fingers can be placed in a wide variety of positions. Additionally, each finger has two interphalangeal joints, where the phalanges articulate with each other. These are simple hinge-shaped joints that allow flexion and extension only.

Body system:	musculoskeletal system
Location:	between the metacarpals and the phalanges, and between the phalanges themselves
Function:	enable versatility of movement in the hand
Components:	metacarpophalangeal and interphalangeal joints
Related parts:	metacarpals, phalanges

Muscles of the Hand

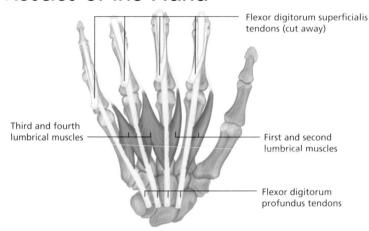

Flexor digitorum superficialis tendons (cut away)

Third and fourth lumbrical muscles

First and second lumbrical muscles

Flexor digitorum profundus tendons

Many of the powerful movements of the hand, which need the contractile strength of a large bulk of muscle tissue, are controlled by the action of muscles in the forearm via tendons. However, precise and delicate actions are produced by small, or "intrinsic" muscles in the hand. These can be divided into three groups: the muscles of the thenar eminence (the bulge of muscle that lies between the base of the thumb and the wrist), which moves the thumb; the muscles of the hypothenar eminence (between the little finger and the wrist); and the short muscles that run deep in the palm of the hand. There are also two groups of muscles that run longitudinally deep within the hand, the lumbricals and the interossei.

Body system:	musculoskeletal system
Location:	within the hand
Function:	enable the hand to make precise and delicate movements
Components:	thenar eminence, hypothenar eminence, lumbricals, interossei
Related parts:	bones of the hand, tendons

Interosseous Muscles

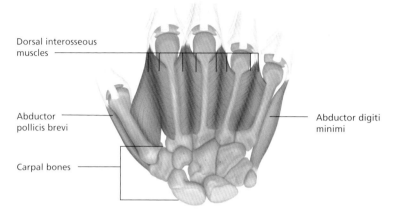

Dorsal interosseous muscles

Abductor pollicis brevi

Carpal bones

Abductor digiti minimi

The interosseous muscles of the hand lie in two layers, those near the palm, the palmar interossei, and the deeper layer, the dorsal interossei. The palmar interossei are small muscles that arise from the palmar surface of the metacarpals (excluding the third). The first two pass around the medial side of each digit before inserting into the dorsal (back) surface. Those that pass to digits four and five pass around the lateral (thumb) side. Contraction of these muscles pulls the fingers together to give the action of adduction. The dorsal interossei muscles are larger and lie between the metacarpal bones, deep to the palmar interossei. Each arises from the sides of the metacarpal bones adjacent to it and acts to spread the fingers.

Body system:	musculoskeletal system
Location:	between the metacarpal bones in the hand
Function:	adduct and abduct the fingers and thumb
Components:	palmar interossei, dorsal interossei
Related parts:	other muscles and tendons in the hand

Thenar and Hypothenar Eminence

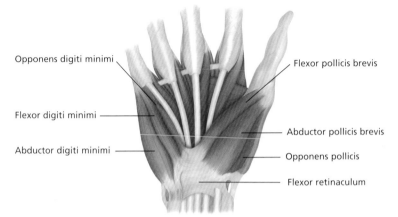

Opponens digiti minimi

Flexor digiti minimi

Abductor digiti minimi

Flexor pollicis brevis

Abductor pollicis brevis

Opponens pollicis

Flexor retinaculum

The muscles that move the thumb are contained in the thenar eminence at its base. Those that move the little finger are found in the hypothenar eminence between the little finger and the wrist. The four small muscles of the thenar eminence act together to adduct, abduct, and flex the thumb and to perform the movement known as "opposition," in which the tip of the thumb is brought into contact with the tips of the fingers. The muscles of the smaller hypothenar eminence form the swelling that lies between the little finger and the wrist. These muscles act together to move the little finger around toward the thumb during the action of cupping the hand or when gripping the lid of a jar to twist it off.

Body system:	musculoskeletal system
Location:	between the thumb and the wrist, and the little finger and the wrist
Function:	enable a range of movements in the thumb and little finger
Components:	abductor pollicis brevis, flexor pollicis brevis, opponens pollicis, flexor retinaculum, abductor digiti minimi, flexor digiti minimi, opponens digiti minimi, palmaris brevis
Related parts:	carpals, metacarpals, phalanges

Soft Tissues of the Hand

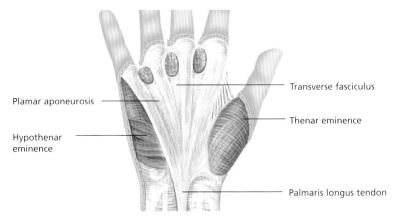

Plamar aponeurosis

Hypothenar eminence

Transverse fasciculus

Thenar eminence

Palmaris longus tendon

The soft tissues of the hand and fingers have an important functional role to play in the hand's actions. The skin of the palm of the hand, and especially the fingers, is attached by fibrous bands to the underlying bones and other tissues. This helps the hand grip efficiently by preventing movement of the skin against these structures. The skin of the back of the hand, which is not involved in gripping or holding, is more mobile. Many of the actions of the hand are concerned with its function as an organ of touch. Numerous nerve endings within the skin and soft tissues of the hand, especially the tips of the fingers, allow a great deal of information to be gathered and sent to the brain for processing.

Body system:	musculoskeletal system
Location:	surrounding the structures within the hand
Function:	bind structures in the hand to the skin to prevent skin from moving freely; help stabilize and anchor structures
Components:	fascia, fibrous bands
Related parts:	skin, structures within the hand

Arteries of the Hand

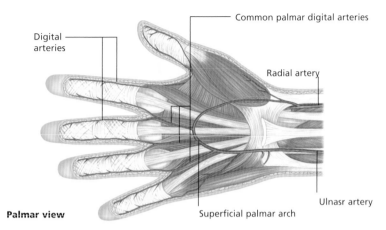

Common palmar digital arteries

Digital arteries

Radial artery

Ulnasr artery

Superficial palmar arch

Palmar view

The hand has a plentiful blood supply from the ulnar and radial arteries. These have many interconnections, maintaining the blood supply even if one artery is damaged. The ulnar artery enters the hand on the side of the little finger and crosses the palm to join with the radial artery to form the "superficial palmar arch." From it arises the small digital arteries that supply blood to the little, ring, and middle fingers. The deep palmar arch is formed by a continuation of the radial artery. This enters the palm from below the base of the thumb and branches into small arteries that supply the thumb and index finger. A network of small arteries lies over the back of the wrist and supplies the back of the hand and fingers.

Body system:	cardiovascular system
Location:	throughout the tissues of the hand
Function:	provide a constant supply of arterial blood, rich in oxygen and nutrients, to the tissues of the hand
Components:	superficial palmar arch, deep palmar arch, digital arteries
Related parts:	ulnar and radial arteries, veins of the hand

Nerves of the Hand

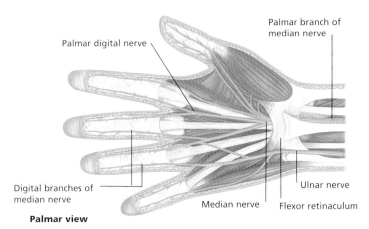

Palmar branch of
median nerve

Palmar digital nerve

Digital branches of
median nerve

Ulnar nerve

Median nerve   Flexor retinaculum

Palmar view

The structures of the hand receive their nerve supply from terminal branches
of the three main nerves in the upper limb: the median, ulnar, and radial (not
shown) nerves. The median nerve enters the hand on the palmar side by passing
under the flexor retinaculum (a restraining band of connective tissue) within
the carpal tunnel. Branches of this nerve supply many of the muscles in the hand,
as well as skin. The ulnar nerve passes over the flexor retinaculum on the medial
side of the hand and its branches also supply various muscles and areas of skin.
The radial nerve runs down the back of the forearm to the dorsal surface (back)
of the hand to supply the skin on the back of three of the fingers.

Body system:	nervous system
Location:	within the hand
Function:	provide motor and sensory nerve supply to the muscles and skin of the hand
Components:	terminal branches of the radial, ulnar, and median nerves
Related parts:	structures and tissues of the hand

Overview of the Abdomen

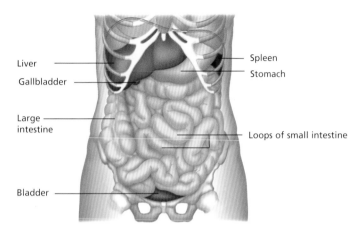

Liver

Gallbladder

Large intestine

Bladder

Spleen

Stomach

Loops of small intestine

The abdomen is the anatomical area that lies between the thorax and the pelvis. The organs of the upper part of the abdominal cavity—the liver, gallbladder, stomach and spleen—lie under the dome of the diaphragm and are protected by the lower ribs. The vertebrae and their associated muscles form the back wall of the abdominal cavity, while the bones of the pelvis support it from beneath. The abdomen is relatively unprotected by bone. This does, however, allow for mobility of the trunk and enables the abdomen to distend when necessary, such as after a meal or during pregnancy. The abdominal cavity contains many organs, including much of the gastrointestinal tract and the kidneys, liver, spleen and gallbladder.

Body system:	contains organs from various systems
Location:	between the thorax and the pelvis
Function:	various functions, depending on organ
Components:	most of the gastrointestinal tract, kidneys, liver, spleen, gallbladder, blood vessels, lymphatics, and nerves, fatty tissue
Related parts:	thorax, pelvis

The Omentum

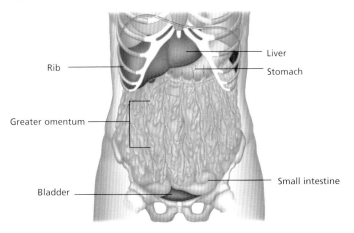

Rib

Liver

Stomach

Greater omentum

Bladder

Small intestine

The majority of the contents of the abdominal cavity are covered in a thin, lubricating sheet of tissue, known as the peritoneum. Folds of the peritoneum attach the abdominal organs to the walls of the abdominal cavity, and enable them to slide easily over one another. The most noticeable part of the peritoneum is the greater omentum, which hangs down from the lower border of the stomach and covers the transverse colon and the coils of small bowel like an apron. The greater omentum contains a large amount of fat, which gives it a yellowish appearance. The greater omentum has been called "the abdominal policeman" due to its action in wrapping itself around an inflamed organ to prevent the spread of infection.

Body system:	digestive system
Location:	hangs down from the lower border of the stomach, covering the transverse colon and small intestine
Function:	protects abdominal organs from injury and infection; insulates abdomen against heat loss
Components:	membrane, fat
Related parts:	rest of peritoneum, abdominal organs

Abdominal Wall

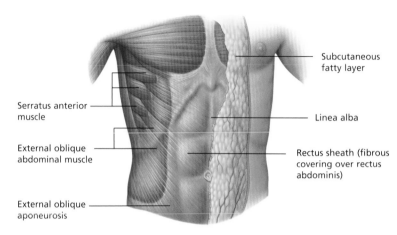

Subcutaneous fatty layer

Serratus anterior muscle

Linea alba

External oblique abdominal muscle

Rectus sheath (fibrous covering over rectus abdominis)

External oblique aponeurosis

The posterior (rear) abdominal wall is formed by the lower ribs, the spine, and accompanying muscles, while the anterolateral (front and sides) wall consists entirely of muscle and fibrous sheets, known as aponeuroses. Under the skin and subcutaneous fat layer lie the muscle layers of the abdominal wall. The muscles here lie in three broad sheets: the external oblique, the internal oblique, and the transversus abdominis, which give the abdomen support in all directions. In addition, there is a wide band of muscle, the rectus abdominis, which runs vertically down the front of the abdomen. The innermost layer of the abdominal wall is the peritoneum, a thin membrane that lines the abdominal cavity.

Body system:	integumentary and musculoskeletal system
Location:	surrounding the abdominal contents, below the rib cage and above the pelvis
Function:	provides support for the abdominal organs; muscles enable flexibility of movement
Components:	skin, superficial fatty layer, superficial membranous layer, three muscle layers, deep fascia layers, transversalis fascia, fat, peritoneum
Related parts:	abdominal organs, thorax, pelvis

Deep Muscles of the Abdomen

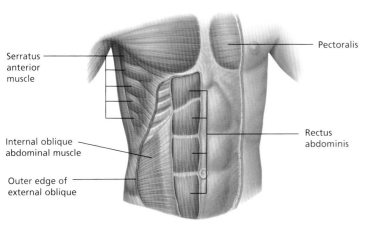

Serratus anterior muscle

Internal oblique abdominal muscle

Outer edge of external oblique

Pectoralis

Rectus abdominis

Beneath the large external oblique muscle lie two more layers of sheetlike muscle, the internal oblique and the transversus abdominis (not shown). Running vertically down the center of the abdominal wall is the rectus abdominis (the "six pack" seen in physically fit people). The internal oblique is a broad thin sheet that lies deep to the external oblique. Its fibers run upward and inward at approximately 90 degrees to the external oblique. The transverse abdominis is the innermost of the three sheets of muscle and supports the stomach. The rectus abdominis is formed of two straplike muscles that run from the front of the rib cage to the pelvis. Lying between each muscle is a thin tendinous band of connective tissue, the linea alba.

Body system:	musculoskeletal system
Location:	in the abdominal wall
Function:	provide support for the abdominal organs; muscles enable flexibility of movement
Components:	internal oblique, transversus abdominis, and rectus abdominis muscles
Related parts:	abdominal organs, muscles of the back

The Stomach

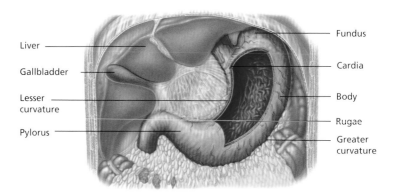

Liver

Gallbladder

Lesser curvature

Pylorus

Fundus

Cardia

Body

Rugae

Greater curvature

The stomach is a distensible, muscular bag lined with mucous membrane, which is positioned between the lower end of the esophagus and the upper end of the small intestine. When empty, the stomach lining lies in numerous folds called rugae, which allow for expansion as it fills with food. The stomach is lined with gastric epithelium and contains numerous glands, and between them they secrete protective mucus, acid, and enzymes. The muscle layer of the stomach wall has oblique, longitudinal, and circular fibers, and this arrangement enables the thorough churning of food. Anatomically, the stomach is said to have four parts, the cardia, fundus, body, and the pylorus, and two curvatures (greater and lesser).

Body system:	digestive system
Location:	between the esophagus and small intestine
Function:	turns solid food into semiliquids by churning it before propelling it onward toward the intestine; destroys bacteria in food; starts the process of digestion
Components:	cardia, fundus, body, pylorus
Related parts:	esophagus, duodenum

The Gastroesophageal Junction

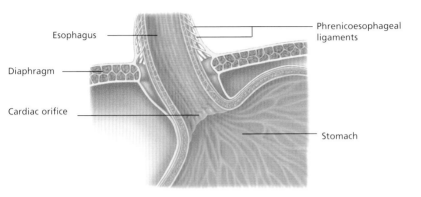

Esophagus

Phrenicoesophageal ligaments

Diaphragm

Cardiac orifice

Stomach

The muscular tube of the esophagus becomes continuous with the stomach just below the diaphragm. There is no identifiable valve at the top of the stomach to control the passage of food. However, it seems that the surrounding muscle fibers of the diaphragm act to keep the tube closed except when a bolus (swallowed mass) of food passes through. The esophagus and upper part of the stomach are held to the diaphragm by the phrenicoesophageal ligaments. These ligaments are extensions of the fascia, a connective tissue that covers the surface of the diaphragm. At the lower end of the esophagus, the epithelium (or lining of cells) changes from multilayered stratified squamous to the typical mucosa present in the stomach.

Body system:	digestive system
Location:	at the junction of the esophagus and stomach
Function:	allows food to pass through from the esophagus to the stomach; prevents backflow of stomach contents
Components:	lower end of esophagus, upper part of stomach, diaphragm
Related parts:	phrenicoesophageal ligaments

Blood Supply to the Stomach

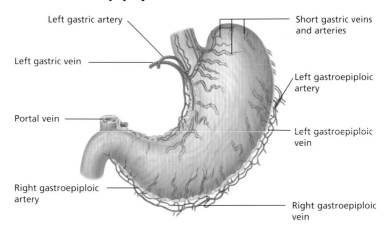

Left gastric artery

Short gastric veins
and arteries

Left gastric vein

Left gastroepiploic
artery

Portal vein

Left gastroepiploic
vein

Right gastroepiploic
artery

Right gastroepiploic
vein

The stomach has a rich blood supply, which comes from the various branches of the celiac trunk, itself a branch of the aorta. The left gastric artery runs along the lesser curvature of the stomach, eventually merging with the terminal branches of the right gastric artery. The splenic artery (not shown), which supplies the spleen, also has major arteries branching off to the greater curvature of the stomach—the short gastric arteries that deliver oxygenated blood to the fundus (top) of the stomach and the left gastroepiploic artery. All the venous blood from the stomach is eventually drained via the veins into the portal venous system, which takes blood to the liver for processing before returning it to the heart.

Body system:	cardiovascular system
Location:	surrounding the tissues of the stomach
Function:	arteries provide blood rich in oxygen and nutrients to the stomach's tissues; veins drain deoxygenated blood and return it to the heart
Components:	left gastric artery and vein, right and left gastroepiploic arteries and veins, short gastric arteries and veins
Related parts:	portal vein, celiac artery, aorta

Lymphatic Drainage of the Stomach

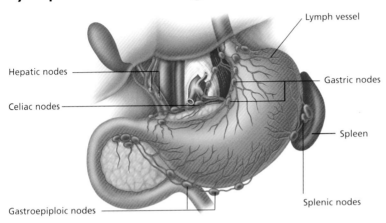

Lymph vessel

Hepatic nodes

Gastric nodes

Celiac nodes

Spleen

Splenic nodes

Gastroepiploic nodes

The lymph vessels and nodes that make up the lymphatic drainage of the stomach follow the general pattern of the arterial blood supply. Lymph from the area supplied by the left and right gastric arteries drains into the left and right gastric lymph nodes, which lie along the lesser curve of the stomach. The splenic nodes are located at the hilum (hollow) of the spleen on the left side of the stomach. These receive lymph from the area supplied by the short gastric arteries. The left and right gastroepiploic nodes lie along the greater curve of the stomach and receive lymph from all areas supplied by the corresponding gastroepiploic arteries. From all these groups of nodes, the lymph travels on to drain into the celiac nodes.

Body system:	lymphatic system
Location:	surrounding the stomach in the same pattern as the arteries
Function:	drain excess fluid from the spaces between the cells that form the stomach's tissues
Components:	lymph nodes and vessels
Related parts:	venous system

The Duodenum

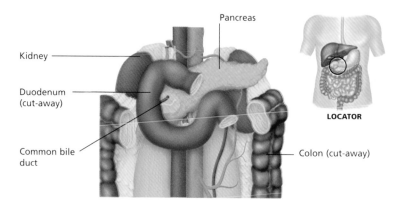

Pancreas

Kidney

Duodenum
(cut-away)

LOCATOR

Common bile
duct

Colon (cut-away)

The small intestine is the main site of food digestion and absorption and is divided into three parts: the duodenum, the jejunum, and the ileum. The duodenum is the first section and the shortest, measuring about 10 in (25 cm) in length. The contents of the stomach pass through the pyloric sphincter and into the duodenum, where they are mixed with secretions from the duodenal wall and from the pancreas and gallbladder. The walls of the duodenum have two layers of muscle fibers, one circular and one longitudinal. The mucosa, or lining, of the walls is particularly thick and contains glands (Brunner's glands) that secrete a thick alkaline fluid, which helps counteract the acidic nature of the stomach contents.

Body system:	digestive system
Location:	between the pylorus of the stomach and the jejunum
Function:	receives stomach contents and mixes them with digestive enzymes and bile; pushes food down the gastrointestinal tract by peristalsis
Components:	muscular walls, inner layer of glandular mucosa
Related parts:	stomach, jejunum, gallbladder, pancreas

The Jejunum and the Ileum

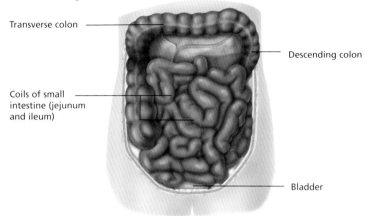

Transverse colon

Descending colon

Coils of small
intestine (jejunum
and ileum)

Bladder

Together, the jejunum and the ileum form the longest part of the small intestine
and are more than 20 ft (6 m) in length. They are surrounded and supported by
a fan-shaped fold of the peritoneum—the mesentery—which allows them to move
within the abdominal cavity with changes of body posture. Food passes from the
duodenum into the jejunum, where most of the nutrients are absorbed. The lining of
the jejunum is thicker than the ileum and contains more folds, or plicae, which slow
the passage of its contents and increase the surface area for absorption (this surface
area is about three times that of the human body). The small intestine has a plentiful
blood supply to enable nutrients and water to pass into the bloodstream.

Body system:	digestive system
Location:	between the duodenum and the cecum (the first part of the large intestine)
Function:	continue the process of digestion and absorption
Components:	jejunum, ileum
Related parts:	stomach, duodenum, cecum

Lining of the Small Intestine

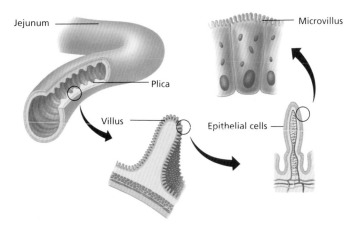

Jejunum

Microvillus

Plica

Villus

Epithelial cells

The small intestine is the main site of absorption of nutrients from ingested food and, although its length alone gives it a large surface area, more is needed for maximum absorption. To achieve this, the lining is structured in circular folds, or plicae, about ⅜ in (1 cm) in depth, to increase the surface area threefold. In addition, tiny fingerlike projections called villi arise from the plicae. About ¹⁄₂₄ in (1 mm) tall, they give a velvety appearance to the lining of the bowel and increase the area for absorption by another 10 times. A further adaptation of the intestinal lining is the presence of microvilli, microscopic hairlike structures (3,000–6,000 on each epithelial cell) projecting from the villi, which increase the surface area still further.

Body system:	digestive system
Location:	lining the small intestine
Function:	to provide the biggest surface area possible for the absorption of nutrients
Components:	plicae, villi, microvilli
Related parts:	blood circulation, stomach, colon

The Cecum

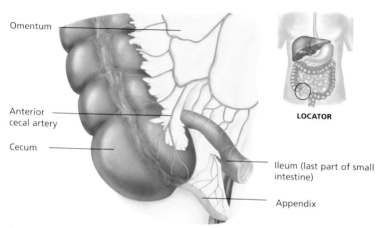

Omentum

Anterior
cecal artery

Cecum

LOCATOR

Ileum (last part of small
intestine)

Appendix

The cecum is a blind-ended pouch about 3 in (75 mm) long at the junction of the small and large intestines. This junction, which is located on the right side of the lower abdomen, is also known as the ileocecal region. Between the cecum and the ileum (the last part of the small intestine) is the ileocecal valve, a ring of circular fibers that opens to allow liquefied food to pass through. From the cecum onward, as far as the sigmoid colon, the external muscular coat of the intestine is separated into three narrow bands, the teniae coli. The arterial blood supply to the cecum is from the anterior and posterior cecal arteries. Venous blood returns through a similar layout of veins, ultimately draining into the superior mesenteric vein.

Body system:	digestive system
Location:	at the junction of the small and large intestines
Function:	absorbs fluid from liquefied feces
Components:	ileocecal valve, muscular coat, mucous lining
Related parts:	ileum, colon, appendix

The Appendix

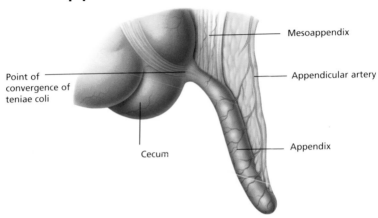

Mesoappendix

Appendicular artery

Point of convergence of teniae coli

Cecum

Appendix

The vermiform (or "wormlike") appendix is attached to the lower end of the cecum, just below the ileocecal valve, and has a free end that extends in one of five directions. It is enclosed within a covering of peritoneum, known as the mesoappendix, which forms a fold between the ileum, the cecum, and the first part of the appendix. The appendix has a complete layer of longitudinal muscle, unlike the rest of the intestine. This is because the three bands of muscle, the teniae coli, converge at the appendix and cover it. The walls of the appendix contain lymphoid tissue, which it is thought may play some role in protection against micro-organisms (such as bacteria), although no real function has been established.

Body system :	digestive system
Location:	extends from the cecum, just below the ileocecal valve
Function:	inconclusive; possibly helps fight against infection
Components:	muscle layer, lymphoid tissue
Related parts:	cecum

The Colon

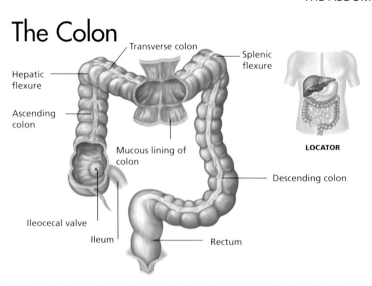

Transverse colon

Splenic flexure

Hepatic flexure

Ascending colon

Mucous lining of colon

LOCATOR

Descending colon

Ileocecal valve

Ileum

Rectum

The colon forms the main part of the large intestine and is about 5 ft (1.5 m) in length. It consists of four sections, which succeed one another in an arch around the abdominal cavity. Liquefied food enters the colon from the small intestine, where it becomes semisolid due to the efficient reabsorption of water through the intestinal walls. There are two sharp bends, or flexures, in the colon known as the hepatic and splenic flexures. The ascending colon runs from the ileocecal valve to the hepatic flexure, where it becomes the transverse colon. At the splenic flexure the next section, the descending colon, runs down to the brink of the pelvis, where it becomes the sigmoid colon, which stores feces prior to defecation.

Body system:	digestive system
Location:	runs between the cecum and the rectum
Function:	absorbs water from liquefied feces
Components:	ascending, transverse, descending, sigmoid colon
Related parts:	small intestine, rectum

Arteries of the Colon

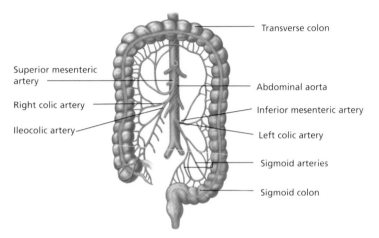

Transverse colon

Superior mesenteric artery

Abdominal aorta

Right colic artery

Inferior mesenteric artery

Ileocolic artery

Left colic artery

Sigmoid arteries

Sigmoid colon

Like the rest of the intestine, each of the sections of the colon is readily supplied with blood from a network of arteries. The arterial supply to the colon comes from the superior and inferior mesenteric branches of the aorta, the large central artery of the abdomen. The ascending colon and first two-thirds of the transverse colon are supplied by the superior mesenteric artery, while the last third of the transverse colon, the descending colon and the sigmoid colon are supplied by the inferior mesenteric. As in other parts of the gastrointestinal tract, there are numerous connections, or anastamoses, between the branches of these two major arteries. In this way an "arcade" of arteries is formed around the wall of the colon.

Body system:	cardiovascular system
Location:	surrounding the colon
Function:	to provide a constant supply of arterial blood rich in oxygen and nutrients to the tissues of the colon
Components:	superior and inferior mesenteric, ileocolic, colic, sigmoid arteries
Related parts:	aorta

Veins of the Colon

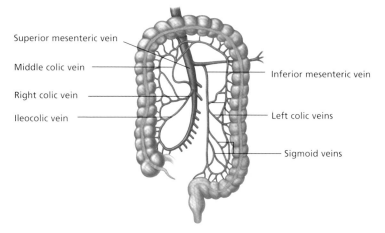

Superior mesenteric vein

Middle colic vein

Right colic vein

Ileocolic vein

Inferior mesenteric vein

Left colic veins

Sigmoid veins

Venous blood from the colon drains ultimately into the portal vein and is carried to the liver, where many of the nutrients absorbed from the intestines are stored or processed. The venous drainage of the colon mirrors the pattern of the arteries. In general, blood from the ascending colon and the first two-thirds of the transverse colon runs into the superior mesenteric vein, with blood from the remainder of the colon being drained by the inferior mesenteric vein. The inferior mesenteric vein drains into the splenic vein, which then joins with the superior mesenteric vein to form the portal vein. Venous blood is transported to the liver in the portal vein, from where it returns to the heart via the inferior vena cava.

Body system:	cardiovascular system
Location:	surrounding the colon
Function:	to drain venous blood from the colon and return it to the heart via the liver
Components:	superior and inferior mesenteric, ileocolic, colic, sigmoid veins
Related parts:	portal vein

The Rectum and Anal Canal

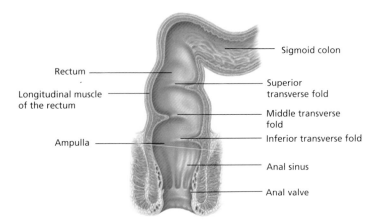

Rectum

Longitudinal muscle
of the rectum

Ampulla

Sigmoid colon

Superior
transverse fold

Middle transverse
fold

Inferior transverse fold

Anal sinus

Anal valve

The rectum and anal canal together form the last part of the gastrointestinal tract. They receive waste matter in the form of feces and enable it to be passed out of the body. The rectum is situated between the sigmoid colon and the anal canal and temporarily stores feces. The longitudinal muscle of the rectum is in two broad bands, which run down its front and back surfaces. There are three horizontal folds in the wall of the rectum, know as the superior, middle, and inferior transverse folds. Below the inferior fold, the rectum widens at the ampulla and then opens into the anal canal, which produces mucus to act as a lubricant. Except during defecation, the anal canal is empty and the sphincter is closed.

Body system:	digestive system
Location:	the final section of the gastrointestinal tract
Function:	rectum stores feces until defecation occurs; anal canal produces lubricating mucus and controls release of feces from the body
Components:	muscle tissue, ampulla, superior, inferior and transverse folds, anal sinuses, valves
Related parts:	anal sphincter, sigmoid colon

The Anal Sphincter

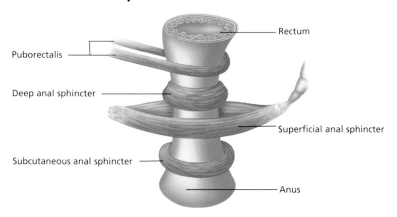

Rectum

Puborectalis

Deep anal sphincter

Superficial anal sphincter

Subcutaneous anal sphincter

Anus

The contents of the intestines move down the digestive tract under "involuntary" control until they reach the rectum. From this stage, further passage of feces is controlled by the various parts of the anal sphincter, which regulates their elimination from the body. The internal anal sphincter is a thickening of the normal circular muscle layer of the bowel in the upper two-thirds of the anal canal; it is not under voluntary control. The puborectalis is a sling of muscle that loops around the anorectal junction, forming an angle and preventing passage of the contents of the rectum into the anal canal. The external anal sphincter (deep, superficial, and subcutaneous) is under voluntary control, and can be relaxed when convenient.

Body system:	digestive system
Location:	at the lower end of the anal canal
Function:	controls release of feces from the body
Components:	muscle tissue, puborectalis, deep, superficial, and subcutaneous anal sphincters
Related parts:	rectum

Veins of the Rectum and Anus

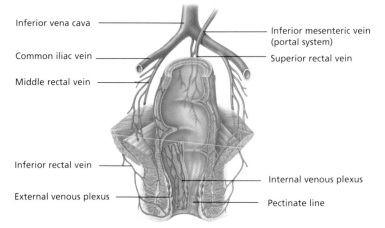

Inferior vena cava

Common iliac vein

Middle rectal vein

Inferior rectal vein

External venous plexus

Inferior mesenteric vein (portal system)

Superior rectal vein

Internal venous plexus

Pectinate line

Beneath the lining of the rectum and anal canal lies a network of small veins, the rectal venous plexus. The plexus is formed of two parts: the internal venous plexus, which lies just under the lining, and the external venous plexus, which lies outside the muscle layer. These two networks receive blood from the tissues and carry it to the larger veins that drain the area—the superior, middle, and inferior rectal veins. The internal venous plexus of the anal canal drains blood in two directions on either side of the pectinate line region. Above this level, blood drains mainly into the superior rectal vein and from there to the portal venous system, while from below it blood drains into the inferior rectal vein.

Body system:	cardiovascular system
Location:	surrounding the rectum and anus
Function:	to drain deoxygenated blood from the rectum and anus and carry it to the liver or return it to the heart
Components:	internal and external venous plexus, inferior, middle and superior rectal veins
Related parts:	common iliac vein, inferior mesenteric vein, inferior vena cava

Nerves of the Rectum and Anus

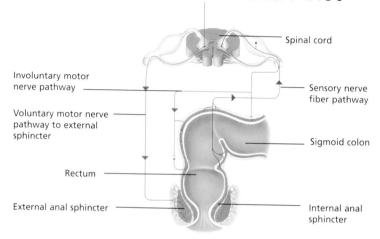

Spinal cord

Involuntary motor nerve pathway

Sensory nerve fiber pathway

Voluntary motor nerve pathway to external sphincter

Sigmoid colon

Rectum

External anal sphincter

Internal anal sphincter

Like the rest of the gastrointestinal tract, the walls of the rectum and anal canal have a nerve supply from the body's autonomic nervous system. This system works "in the background," usually without us being aware of it, to regulate and control the body's internal functions. When the rectum fills with feces, nerve endings trigger an involuntary reflex contraction of the walls of the rectum, and feces enter the anal canal. However the anal canal, or more specifically the external anal sphincter, has a nerve supply from the "voluntary" nervous system. These nerves, which originate in the sacral spinal nerves, allow us to contract and relax the sphincter at will, and therefore to control whether or not to defecate.

Body system:	nervous system
Location:	surrounding the rectum and anus
Function:	to enable the involuntary contraction and relaxation of the walls of the rectum and the voluntary control of the anal sphincter
Components:	parasympathetic nerve fibers, spinal nerves
Related parts:	spinal cord

Lymph Drainage of the Intestines

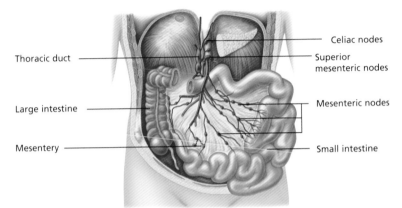

Celiac nodes

Thoracic duct

Superior mesenteric nodes

Mesenteric nodes

Large intestine

Mesentery

Small intestine

Many of the lymph nodes that drain the intestines lie within the mesentery, a fold of tissue that binds the gut to the abdominal wall. Nodes are found in a number of areas: by the wall of the intestine, running close to the arteries, and alongside the large superior and inferior mesenteric arteries. These mesenteric groups of nodes are, in some cases, named according to their positions in relation to the intestine or to the artery that they accompany. From the intestinal wall, lymph drains through these nodes to eventually reach the pre-aortic nodes that lie next to the aorta. In addition to its normal function, the lymph that leaves the small intestine also transports the fats absorbed from food to the bloodstream.

Body system:	lymphatic system
Location:	within the mesentery and surrounding the intestines
Function:	drains excess fluid from around the cells of the intestines and returns it to the bloodstream; acts as a filter for harmful organisms; transports absorbed fat and delivers it to the bloodstream
Components:	lymph vessels, mesenteric nodes, celiac nodes, pre-aortic nodes, etc.
Related parts:	mesentery, vascular system, rest of lymphatic system

Microscopic Anatomy

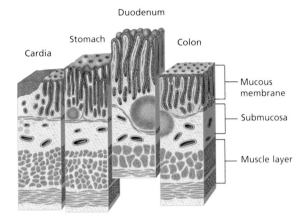

Cardia · Stomach · Duodenum · Colon

- Mucous membrane
- Submucosa
- Muscle layer

The structure of the walls of the digestive tract varies according to the function of the particular area. The entire digestive tract is lined with mucous membrane, which contains glands that secrete mucus to help the smooth passage of food through the system. In addition, the lining of the stomach and small intestine has glands that produce digestive juices. Most of the absorption of nutrients takes place in the small intestine (duodenum, jejunum, and ileum) and here multiple projections called villi increase the surface area for absorption by many times. The more flattened surface of the colon (large intestine) is ideal for the absorption of water, which is its primary function, and is pitted with mucus-secreting glands.

Body system:	digestive system
Location:	throughout the gastrointestinal tract
Function:	breakdown and digestion of food, absorption of nutrients and water, excretion of waste products
Components:	esophagus, stomach, intestines
Related parts:	vascular system, liver, pancreas, gallbladder

The Liver

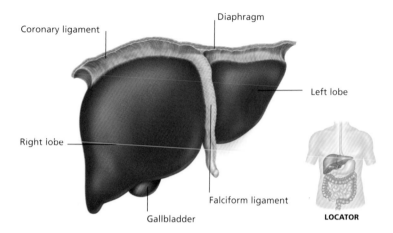

Coronary ligament

Diaphragm

Left lobe

Right lobe

Falciform ligament

Gallbladder

LOCATOR

The liver is the largest abdominal organ, weighing about 3½ lb (1.5 kg) in adult men. It has many important metabolic and digestive functions, and also produces bile, which is stored in the gallbladder. The liver lies under the diaphragm on the right of the abdominal cavity and is protected by the rib cage. Although it has four lobes, functionally the liver is divided into two parts, right and left, each receiving its own rich blood supply. The two smaller lobes, the caudate and quadrate, can only be seen on the underside of the liver. The greater part of the liver is covered with the peritoneum, a sheet of connective tissue lining the walls and structures of the abdomen. Folds of the peritoneum form the various ligaments of the liver.

Body system:	digestive system
Location:	under the diaphragm on the right side of the abdomen
Function:	many metabolic functions; plays major role in processing nutrients and destroying toxins; produces bile, which breaks down fat
Components:	right, left, quadrate, and caudate lobes
Related parts:	biliary system, portal and systemic venous systems

Microscopic Anatomy of the Liver

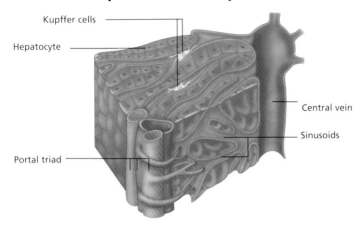

Kupffer cells

Hepatocyte

Central vein

Sinusoids

Portal triad

The liver is composed of numerous tiny groups of cells called lobules. These have a distinctive hexagonal-shaped structure, with hepatocytes (liver cells) arranged like the spokes of a wheel around a central vein—a tributary of the hepatic vein. Blood flows past the hepatocytes and into this central vein through tiny vessels known as sinusoids, which receive blood from the portal triads, groupings of three vessels that lie at the six points of the lobule. The portal triad consists of a branch of the hepatic artery and portal vein, and a biliary duct that collects bile made by the hepatocytes. The sinusoids contain tiny specialized Kupffer cells, which remove debris and worn-out cells from the blood before it returns to the heart.

Body system:	digestive system
Location:	under the diaphragm on the right side of the abdomen
Function:	many metabolic functions; plays major role in processing nutrients and destroying toxins; produces bile, which breaks down fat
Components:	hepatocytes, sinusoids, portal triads, Kupffer cells
Related parts:	biliary system, portal and systemic venous systems

Visceral Surface of the Liver

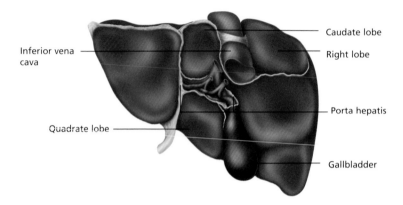

Caudate lobe

Inferior vena cava

Right lobe

Porta hepatis

Quadrate lobe

Gallbladder

The underside of the liver is known as the visceral surface as it lies against the abdominal organs, or viscera, such as the kidney and intestines. Because the tissue of the liver is soft and pliable, these surrounding structures may leave impressions on its surface. The two smaller lobes of the liver—the caudate lobe and the quadrate lobe—and the gallbladder are also visible on the visceral surface. In the center of the liver is the porta hepatis, an area that is similar to the hilum of the lungs; from this area, major vessels enter and leave the liver clothed in a sleeve of peritoneum. Structures that pass through the porta hepatis include the portal vein, the hepatic artery, the bile ducts, lymphatic vessels, and nerves.

Body system:	digestive system
Location:	under the diaphragm on the right side of the abdomen
Function:	many metabolic functions; plays major role in processing nutrients and destroying toxins; produces bile, which breaks down fat
Components:	right, left, caudate, quadrate lobes, porta hepatis
Related parts:	biliary system, portal and systemic venous systems

The Biliary System

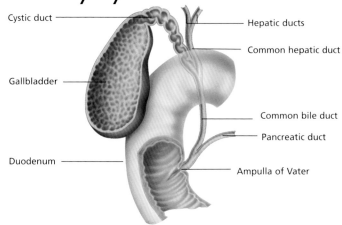

Cystic duct

Hepatic ducts

Common hepatic duct

Gallbladder

Common bile duct

Pancreatic duct

Duodenum

Ampulla of Vater

Bile is a greenish fluid that aids the digestion of fat within the duodenum, the first section of the small intestine. Bile is secreted by the hepatocytes (liver cells) and passes down the right and left hepatic ducts, which leave the liver through the porta hepatis and then merge to form the common hepatic duct. Excess bile is stored and concentrated in a small thin-walled sac on the underside of the liver called the gallbladder. When fatty substances enter the duodenum, the gallbladder contracts and expels the bile. The gallbladder is connected to the common hepatic duct by the cystic duct; these two ducts merge to form the common bile duct, which, along with the pancreatic duct, drains into the duodenum at the ampulla of Vater.

Body system:	digestive system
Location:	ducts pass from the liver down to the duodenum; gallbladder sits just under the visceral surface of the liver
Function:	transports bile from the liver to the small intestine to aid the digestion of fat
Components:	right and left hepatic ducts, common hepatic duct, gallbladder, cystic duct, common bile duct, ampulla of Vater
Related parts:	liver, duodenum, pancreas

The Pancreas

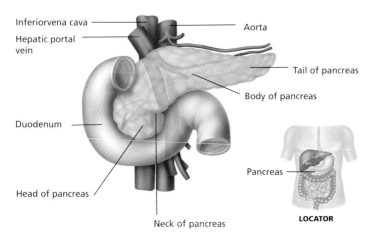

Inferiorvena cava

Hepatic portal vein

Aorta

Tail of pancreas

Body of pancreas

Duodenum

Pancreas

Head of pancreas

Neck of pancreas

LOCATOR

The pancreas is a large pale-colored gland that secretes enzymes into the duodenum, the first part of the small intestine, to aid the digestion of protein, fat and carbohydrates (starch). It also produces the hormones insulin and glucagon, which regulate the use of glucose (sugar) by cells. Lying across the posterior wall of the abdomen just behind the stomach, the pancreas is said to have four parts. The head lies in the C-shaped curve of the duodenum, attached to its inner side. The neck is narrower than the head, due to the large hepatic portal vein behind it. The body is triangular in cross-section and lies in front of the aorta; it passes up and to the left to merge with the tail, which comes to a tapering end.

Body system:	digestive/endocrine systems
Location:	behind the stomach across the posterior wall of the abdomen
Function:	produces digestive enzymes; secretes hormones, glucagon, and insulin, that regulate sugar levels in the body
Components:	head, neck, body, and tail of pancreas
Related parts:	duodenum

The Spleen

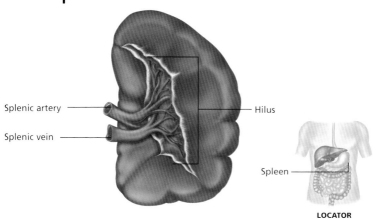

Splenic artery

Splenic vein

Hilus

Spleen

LOCATOR

The largest of the lymphatic organs, the spleen, lies under the lower ribs on the left side of the abdomen. The spleen filters the blood for debris, such as bacteria and worn-out blood cells, and it also produces white blood cells. The hilus (central hollow) of the spleen contains its blood vessels (the splenic artery and vein) and some lymphatic vessels. The spleen is dark purple in color, about the size of a clenched fist, and is surrounded and protected by a thin capsule composed of connective tissue, projections of which pass down into the soft splenic tissue, providing support. Contained within this capsule are muscle fibers that allow the spleen to contract periodically to expel blood back into the circulation.

Body system:	lymphatic system
Location:	under the lower ribs on the left of the abdomen
Function:	filters blood for foreign material, such as bacteria, destroys worn-out blood cells, produces lymphocytes to help fight infection
Components:	red and white "pulp" surrounded by a fibrous capsule
Related parts:	rest of lymphatic system, blood circulation

The Urinary Tract

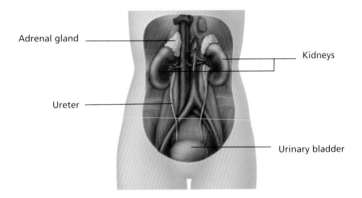

Adrenal gland

Kidneys

Ureter

Urinary bladder

The organs of the urinary tract are together responsible for the production of urine and its expulsion from the body. The two bean-shaped kidneys lie within the abdomen, against the posterior abdominal wall and behind the intestines. They filter the blood to remove waste chemicals and excess fluid, which are excreted as urine. Urine flows from the kidneys down the right and left ureters, long narrow tubes that actively propel the fluid down toward the bladder by muscular contractions. The urine is received and stored in the urinary bladder, a muscular, balloonlike structure that lies within the pelvis. When appropriate, the bladder contracts to expel its contents via the urethra, a thin-walled muscular tube.

Body system:	urinary system
Location:	abdomen and pelvis
Function:	filters blood and removes toxins and excess fluid; the resulting urine is expelled from the body
Components:	right and left kidneys and ureters, bladder and urethra
Related parts:	blood circulation

The Adrenal Glands

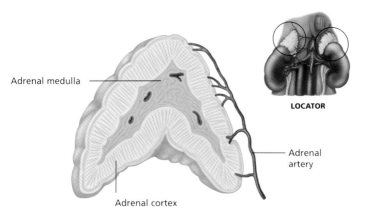

Adrenal medulla

Adrenal cortex

LOCATOR

Adrenal artery

Lying on top of the kidneys are the paired adrenal glands, also known as the suprarenal glands. Although they are physically close to the kidneys, they play no part in the urinary system. Instead, they are endocrine glands that secrete hormones vital to the healthy functioning of the body. The adrenal glands are made up of two parts, a central medulla and an outer region known as the cortex, each of which has a separate function. The yellow adrenal cortex makes up the bulk of the gland. It secretes a wide variety of hormones collectively known as corticosteroids. The darker adrenal medulla is formed of a "knot" of nervous tissue surrounded by blood vessels and is the site of production of adrenaline and noradrenaline.

Body system:	endocrine system
Location:	above the kidneys
Function:	produce corticosteroids (vital for the control of metabolism, fluid balance, and response to stress), male sex hormones, and adrenaline and noradrenaline (prepare body for "fight-or-flight" response)
Components:	adrenal cortex and medulla
Related parts:	nervous system

Kidneys

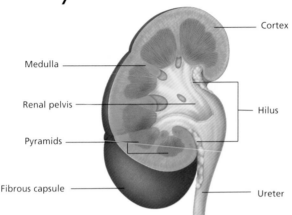

Cortex

Medulla

Renal pelvis

Hilus

Pyramids

Fibrous capsule

Ureter

The paired kidneys lie within the abdominal cavity against the posterior wall. Each kidney is about 4 in (10 cm) in length, reddish brown in color and has a characteristic bean shape. On the inward facing surface lies the hilus of the kidney from which the blood vessels enter and leave. Each kidney is covered with a tough, fibrous capsule and is surrounded by a protective layer of fat. The outer part of the kidney is called the renal cortex and contains nephrons (the functional units of the kidney). The renal medulla is the central area and houses the "pyramids," which consist of the urine-collecting ducts. At the center is the renal pelvis, a funnel-like area in which urine collects before flowing down the ureter.

Body system:	urinary system
Location:	toward the back of the abdominal cavity, just above waist height
Function:	filter blood to remove excess fluid and toxins
Components:	capsule, cortex, medulla, pyramids, pelvis, nephrons
Related parts:	blood circulation, ureters, bladder

Nephrons

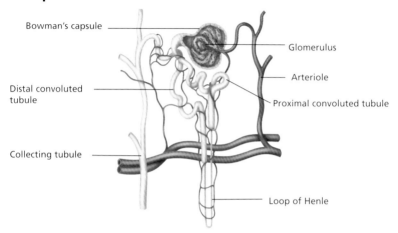

Bowman's capsule

Glomerulus

Arteriole

Distal convoluted tubule

Proximal convoluted tubule

Collecting tubule

Loop of Henle

The work of each kidney is carried out by more than a million tiny filtering units called nephrons. Each of these nephrons consists of a renal corpuscle, from which projects a long loop of tubule. The renal corpuscle is composed of a clump of tiny arterioles (tiny arteries) called the glomerulus and the Bowman's capsule that surrounds it. Fluid and solutes in the blood pass from the blood in the glomerulus to the renal tubule for processing. The renal tubule is a convoluted tube that passes downward and then loops back up again as the Loop of Henle. During the passage through the tubule, most of the solutes and fluid are reabsorbed. The remainder passes down the collecting duct to the pelvis of the kidney and is excreted as urine.

Body system:	urinary system
Location:	within the kidney
Function:	filters arterial blood; fluid and solutes pass into the tubule for either reabsorption into the blood or excretion as urine
Components:	Bowman's capsule, glomerulus, proximal, and distal convoluted tubules, loop of Henle, collecting tubule
Related parts:	blood circulation, other parts of kidney, ureter

Blood Supply to the Kidneys

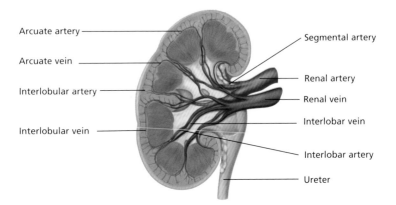

Arcuate artery

Arcuate vein

Interlobular artery

Interlobular vein

Segmental artery

Renal artery

Renal vein

Interlobar vein

Interlobar artery

Ureter

Every day the kidneys process about 450 gal (1,700 L) of blood between them. Arterial blood is carried to each kidney by a renal artery, which arises directly from the aorta, the main artery in the body. The renal artery enters the kidney and divides into segmental arteries, each of which further divides into lobar arteries. The interlobar arteries pass between the renal pyramids and branch to form the arcuate arteries, which run along the junction of the cortex and the medulla. Numerous interlobular arteries pass into the tissue of the renal cortex to carry blood to the nephrons where it is filtered. Venous blood enters the interlobular, arcuate, and then the interlobular veins before being collected by the renal vein.

Body system:	cardiovascular system
Location:	within the kidney
Function:	provide blood rich in oxygen and nutrients to the tissues of the kidney; take arterial blood to the nephrons for filtering; return deoxygenated blood to the venous circulation
Components:	renal artery and vein, segmental arteries, interlobar arteries, etc.
Related parts:	tissues of the kidney

The Bladder and Urethra

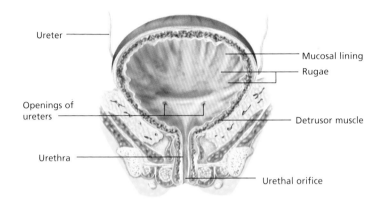

Ureter

Mucosal lining

Rugae

Openings of ureters

Detrusor muscle

Urethra

Urethal orifice

The bladder is a pyramid-shaped sac positioned low in the pelvis just behind the pubic bone, whose function is to store urine. The wall of the bladder has three layers: the innermost layer of mucous membrane is thrown into folds, or rugae, that allow for expansion as the bladder fills. The middle layer, known as the detrusor muscle, consists of smooth muscle fibers running in both circular and longitudinal directions and allows the bladder to contract to expel urine when appropriate. The outer surface is covered by peritoneum over its upper section and by a fibrous adventitia over the remainder. When necessary, urine passes from the bladder down the urethra, a thin-walled muscular tube with a sphincter at its upper end.

Body system:	urinary system
Location:	behind the pubic bone in the pelvis
Function:	stores urine until it is appropriate to urinate; expels urine by muscular contractions
Components:	mucous membrane lining, detrusor muscle, outer layer of peritoneum and fibrous adventitia
Related parts:	kidneys, ureters, pelvic floor muscles

217

Differences in Anatomy

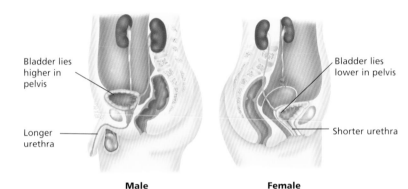

Bladder lies higher in pelvis

Longer urethra

Bladder lies lower in pelvis

Shorter urethra

Male　　　　　**Female**

Owing to the presence of the reproductive organs, the position of the bladder, and the size, shape, and position of the urethra, vary between males and females. The female bladder lies lower in the pelvis, in front of the vagina and uterus and behind the pubic bone, whereas the male bladder lies higher in the pelvis, slightly above the pubic bone. In men, the urethra is approximately 8 in (20 cm) long (five times that of a female) and passes through the prostate gland and down the length of the penis before opening out at the external urethral orifice. In women, the urethra is much shorter—only 1⅛–1⅜ in (3–4 cm) long—and opens at the urethral orifice, which lies just in front of the vaginal opening.

Body system:	urinary system
Location:	pelvis
Function:	bladder acts as a reservoir for urine; urethra allows urine to pass out of the body via the external urethral orifice
Components:	bladder, urethra, external urethral orifice
Related parts:	reproductive organs, pubic bone

The Ureters

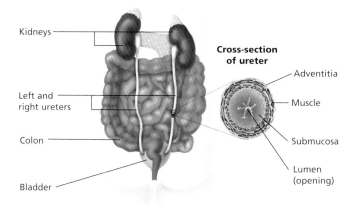

Kidneys

Left and right ureters

Colon

Bladder

Cross-section of ureter

Adventitia

Muscle

Submucosa

Lumen (opening)

The ureters propel urine toward the bladder by peristalsis—muscular contractions of their walls. Each ureter is 9¾–12 in (25–30 cm) in length and ⅛ in (3 mm) wide. They are multilayered structures, consisting of an outer "adventitia" (a protective covering), a muscle layer, submucosal connective tissue, and a lining of urothelium. The ureters start at the kidney and pass down the posterior abdominal wall to cross the bony brim of the pelvis and enter the bladder by piercing its posterior wall. The first funnel-shaped section of the ureter is known as the renal pelvis and lies within the hilus of the kidney. This tapers to form the narrower ureteric tube, which continues downward as the abdominal, then the pelvic ureter.

Body system:	urinary system
Location:	originate at the hilus of the kidney and pass down the abdomen to enter the bladder
Function:	transport urine from the kidney to the bladder
Components:	outer protective covering known as the adventitia, circular and longitudinal layers of muscle, submucosa, lining of urothelium
Related parts:	kidneys, bladder

Bones of the Pelvis

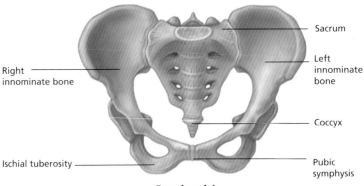

Sacrum

Left innominate bone

Right innominate bone

Coccyx

Ischial tuberosity

Pubic symphysis

Female pelvis

The pelvic bones form a basinlike structure that connects the spine to the lower limbs and protects the contents of the pelvis, including the bladder and the reproductive organs. Many powerful muscles are attached to these bones, allowing the weight of the body to be transferred to the legs with great stability. The bones of the pelvis include the innominate (hip) bones, the sacrum and the coccyx. The innominate bones meet at the pubis symphysis anteriorly, and posteriorly they are joined at the sacrum. Extending down from the sacrum at the back of the pelvis is the coccyx. Male and female pelvises differ in structure, mainly due to the requirements of childbirth but also because men are usually heavier than women.

Body system:	musculoskeletal system
Location:	at the base of the spine at the junction with the lower limbs, encircling the pelvic organs
Function:	protect organs within the pelvis, enable weight of body to be shifted to legs with stability, provide attachments for major muscles
Components:	right and left innominate bones, sacrum, coccyx
Related parts:	spine, bones of lower limbs, pelvic organs

Ligaments of the Pelvis

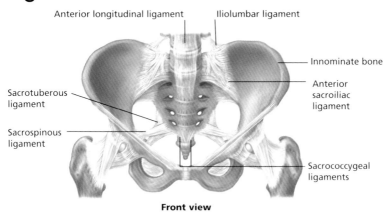

Anterior longitudinal ligament

Iliolumbar ligament

Innominate bone

Anterior sacroiliac ligament

Sacrotuberous ligament

Sacrospinous ligament

Sacrococcygeal ligaments

Front view

The pelvis must be structurally strong in order to perform its functions of transferring weight to the legs and supporting the abdominal contents. The overall stability of the pelvic bones—the innominate (hip) bones, the sacrum, and the coccyx—is helped by a series of tough pelvic ligaments (the pelvic ligaments are among the strongest in the body) that bind the joints between the bones together. These ligaments are generally named after the two areas of bone they link; for example the sacroiliac ligament connects the sacrum and the ilium of the hip bone. The pubic symphysis is the joint between the two pubic bones, which allows almost no movement. It is held in place by the superior and inferior pubic ligaments.

Body system:	musculoskeletal system
Location:	between the bones of the pelvis, and surrounding structures
Function:	produce stability between the bones of the pelvis
Components:	iliolumbar, sacrospinous, sacroiliac, sacrococcygeal, sacrotuberous, anterior longitudinal ligaments
Related parts:	bones and joints of the pelvis and spine

The Hip Bone

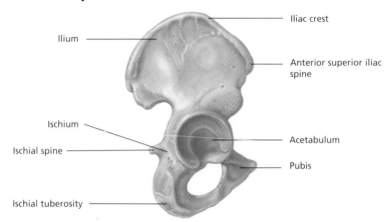

Iliac crest

Ilium

Anterior superior iliac spine

Ischium

Acetabulum

Ischial spine

Pubis

Ischial tuberosity

The two sturdy innominate, or hip, bones constitute the greater part of the pelvis, connecting with each other at the front and with the sacrum at the back. The innominate bone is formed by the fusion of three separate bones: the ilium, the ischium, and the pubis. In children, these bones are joined only by cartilage, but by puberty, they have fused together to form the single innominate bone. The upper margin of the hip bone is formed by the widened iliac crest. Further down the bone is the ischial tuberosity, a large projection of the ischium that bears the body's weight when sitting. The acetabulum is a cuplike socket, which receives the head of the femur (thigh bone) to form the hip joint.

Body system:	musculoskeletal system
Location:	pelvis
Function:	transmits forces between legs and spine and provides stability; helps protect the pelvic organs
Components:	ilium, ischium, pubis
Related parts:	sacrum, femur, muscles and ligaments

The Pelvic Floor Muscles

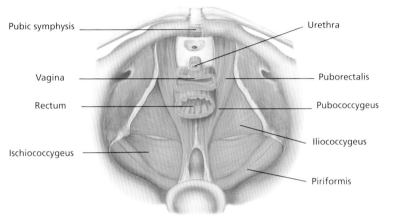

Pubic symphysis

Vagina

Rectum

Ischiococcygeus

Urethra

Puborectalis

Pubococcygeus

Iliococcygeus

Piriformis

The pelvic floor muscles (also known as the pelvic diaphragm) play an important role in supporting the abdominal and pelvic organs. In pregnancy, these muscles help carry the growing weight of the uterus, and during childbirth they support the baby's head as the cervix dilates. The muscles of the pelvic floor are attached to the inside of the ring of bone that makes up the pelvic skeleton, and they slope downward to form a rough funnel shape. The levator ani is the largest muscle of the pelvic floor. It is a wide, thin sheet made up of three parts: the pubococcygeus (the main part), the puborectalis, and the iliococcygeus. A second muscle, the coccygeus (or ischiococcygeus), lies behind the levator ani.

Body system:	musculoskeletal system
Location:	form a muscular floor at the base of the pelvis
Function:	support abdominal and pelvic organs; support growing uterus during pregnancy; assist in the control of defecation and urination
Components:	pubococcygeus, puborectalis, iliococcygeus, coccygeus muscles
Related parts:	pelvic and abdominal organs, pelvic bones and ligaments

Openings of the Pelvic Floor

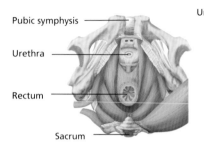

Pubic symphysis

Urethra

Rectum

Sacrum

**Male pelvic diaphragm
from below**

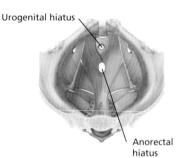

Urogenital hiatus

Anorectal
hiatus

**Male pelvic diaphragm
from above**

The pelvic floor resembles the diaphragm in the chest in that it forms a nearly continuous sheet of muscle. However, it has two openings, or hiatuses, that enable important structures to run through it: the anorectal hiatus and the urogenital hiatus. The anorectal hiatus allows the passage of the rectum and the anal canal from the pelvis to the anus beneath the pelvic floor muscles. The U-shaped fibers of the puborectalis muscle form the posterior edge of this hiatus. Lying in front of the anorectal hiatus is the urogenital hiatus, through which the urethra (which carries urine from the bladder out of the body) passes. In females, the vagina also passes through the pelvic diaphragm within the urogenital hiatus.

Body system:	musculoskeletal system
Location:	form a muscular floor at the base of the pelvis
Function:	enable the rectum, anal canal, urethra, and vagina to pass through the pelvic diaphragm
Components:	anorectal hiatus, urogenital hiatus
Related parts:	pelvic floor muscles

The Inguinal Canal

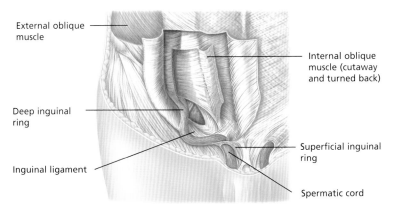

External oblique muscle

Internal oblique muscle (cutaway and turned back)

Deep inguinal ring

Superficial inguinal ring

Inguinal ligament

Spermatic cord

The abdominal wall around the inguinal region, commonly known as the groin, has an area of weakness due the presence of the inguinal canal, through which pass the spermatic cord in males and the round ligament in females. However, the design of the canal minimizes the likelihood of herniation (protrusion) of abdominal contents. The canal passes down toward the midline of the body from its entrance, the deep inguinal ring, to emerge at the superficial inguinal ring. The inguinal canal has a roof, a floor, and two walls formed by abdominal muscles and ligaments. The muscle fibers arching over the canal automatically contract if the abdominal pressure is raised—for example, during coughing or sneezing—to protect its contents.

Body system:	musculoskeletal system
Location:	in the inguinal region, or groin
Function:	provides passageway for spermatic cord and round ligament
Components:	roof formed by internal oblique and transversus abdominis, floor formed by inguinal ligament, anterior wall formed by external oblique, posterior wall formed by transversalis fascia
Related parts:	abdominal muscles, inguinal ligament

The Inguinal Ligament

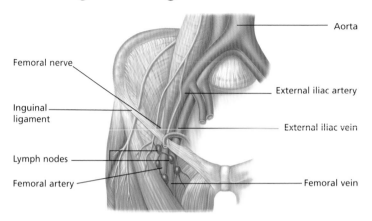

Femoral nerve

Inguinal ligament

Lymph nodes

Femoral artery

Aorta

External iliac artery

External iliac vein

Femoral vein

The inguinal ligament is a tough fibrous band that bridges a gap at the front of the pelvis. Behind it are a number of vital structures, including blood vessels and nerves, that serve the lower limb and two groups of lymph nodes (superficial and deep). Two major blood vessels pass behind the ligament to the lower limb: the femoral artery and the femoral vein. These vessels are enclosed within a thin, funnel-shaped sheet of connective tissue known as the femoral sheath, which allows the vessels to glide harmlessly against the inguinal ligament during hip movements. Lateral to these vessels, on the outer side, lies the femoral nerve, which is the largest branch of the lumbar plexus, a network of nerves within the abdomen.

Body system:	musculoskeletal system
Location:	passes between the anterior superior iliac spine and the pubic tubercle
Function:	forms the floor of the inguinal canal
Components:	formed by the inferior, underturned fibers of the aponeurosis of the external oblique muscle
Related parts:	pelvic bones, abdominal muscles

The Male Reproductive System

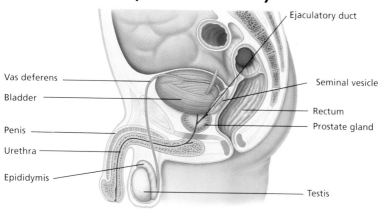

- Ejaculatory duct
- Vas deferens
- Bladder
- Penis
- Urethra
- Epididymis
- Seminal vesicle
- Rectum
- Prostate gland
- Testis

The structures that make up the male reproductive tract are responsible for the production of sperm and seminal fluid and their passage out of the body. Unlike other organs, it is not until puberty that they become fully functional. The paired testes lie suspended in the scrotum outside the body, and they are responsible for the manufacture of sperm, which then travel through a complex network of tubes and ducts. The coiled epididymis receives the sperm from the testis, from where it passes to the vas deferens, a long muscular tube that passes through two glands, the seminal vesicle and the prostate gland, both of which add fluid to the sperm. On leaving the prostate gland, the urethra becomes the central core of the penis.

Body system:	reproductive system
Location:	external genitalia and within the pelvis
Function:	manufacture sperm, produce seminal fluid and enzyme-rich secretions, enable contact between sperm and female egg (ovum)
Components:	testes, scrotum, epididymis, penis, prostate gland
Related parts:	urinary system

Male External Genitalia

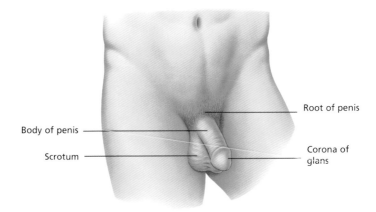

Root of penis

Body of penis

Scrotum

Corona of glans

The external genitalia are those parts of the reproductive tract that are visible in the pubic region, which in a male comprise the scrotum and the penis. The scrotum is a loose bag of skin and connective tissue that holds the paired testes suspended within it. There is a midline septum, or partition, in the scrotum that separates the two testes. Although it seems unusual for the testes to be held in such a vulnerable position outside the protection of the body cavity, a cool temperature is necessary for sperm production. The penis consists mainly of erectile tissue, which becomes engorged with blood during sexual arousal. The urethra, through which urine and sperm pass, runs through the center of the penis.

Body system :	reproductive system
Location:	pubic region
Function:	scrotum keeps testes cool to allow sperm production; penis allows passage of urine and sperm out of the body, enables sexual intercourse
Components:	scrotum, penis
Related parts:	urinary system

The Prostate Gland

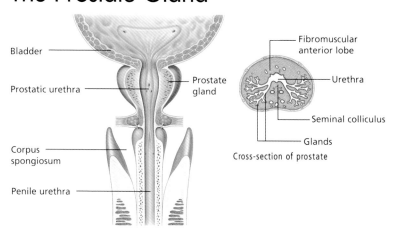

Bladder

Prostatic urethra

Corpus spongiosum

Penile urethra

Prostate gland

Fibromuscular anterior lobe

Urethra

Seminal colliculus

Glands

Cross-section of prostate

About the size of a large walnut and surrounded by a tough fibrous capsule, the prostate gland lies just under the bladder, encircling the upper part of the urethra. The gland is closely attached to the base of the bladder, with its rounded anterior (front) surface lying just behind the pubic bone. The prostate gland forms a vital part of the male reproductive system because it provides an enzyme-rich fluid that helps to activate sperm and forms up to one-third of the total volume of seminal fluid. This fluid is secreted by cells in the prostate gland and passed through ejaculatory ducts, which open into the urethra (called the prostatic urethra at this level) on a raised ridge, the seminal colliculus.

Body system:	reproductive system
Location:	at the base of the bladder surrounding the urethra
Function:	Provides enzyme-rich fluid that provides the bulk of the seminal fluid and helps to activate the sperm
Components:	fibrous capsule, glandular tissue, fibromuscular tissue, ducts
Related parts:	urethra, bladder

The Seminal Vesicles

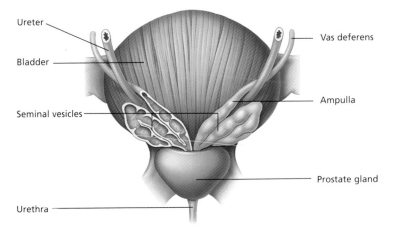

Ureter

Bladder

Seminal vesicles

Urethra

Vas deferens

Ampulla

Prostate gland

The paired seminal vesicles are accessory glands of the male reproductive tract and produce a thick, sugary, alkaline fluid that forms the main part of the seminal fluid. Each seminal vesicle is an elongated structure about the size and shape of the little finger and lies behind the bladder and the front of the rectum, together forming a V shape. Inside the glands are coiled secretory tubules with muscular coats consisting of both circular and longitudinal fibers. The secretions leave the gland in the duct of the seminal vesicle, which usually pierces the capsule of the prostate gland before joining the vasa deferentia to form the ejaculatory duct. In old age, the seminal vesicles shrink to be almost undetectable.

Body system:	reproductive system
Location:	behind the bladder and in front of the rectum
Function:	adds bulk to the seminal fluid, provides an alkaline buffer to the acidity of vaginal secretions
Components:	smooth muscle coat, secretory tubules, ducts
Related parts:	prostate gland, urethra

The Testes and Epididymis

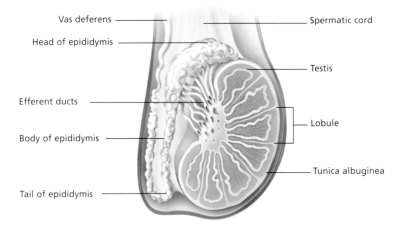

Vas deferens

Head of epididymis

Efferent ducts

Body of epididymis

Tail of epididymis

Spermatic cord

Testis

Lobule

Tunica albuginea

The testes are firm, mobile, oval-shaped structures that are enclosed within a tough protective capsule and suspended from the spermatic cord within the scrotum. Each testis is formed of lobules that contain several tightly coiled seminiferous tubules, the site of sperm production. In the connective tissue surrounding these tubules are the Leydig cells, which produce male hormones. Efferent ducts carry the sperm to the epididymis, a tightly coiled tube about 20 ft (6 m) in length that lies closely attached to the upper part of the testis. It is in the epididymis that the sperm are stored until they mature. Sperm leave the testis via the vas deferens, which extends from the tail of the epididymis.

Body system:	reproductive system
Location:	suspended within the scrotum
Function:	manufacture and store sperm, produce male hormones
Components:	scrotum, testes, efferent ducts, epididymes, vasa deferentia
Related parts:	urinary system

The Scrotum

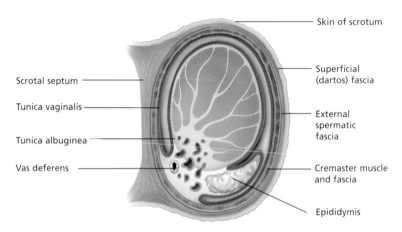

Skin of scrotum

Superficial (dartos) fascia

External spermatic fascia

Cremaster muscle and fascia

Epididymis

Scrotal septum

Tunica vaginalis

Tunica albuginea

Vas deferens

Normal sperm can only be produced if the temperature is about three degrees lower than the internal body temperature. Muscle fibers within the spermatic cord and the walls of the scrotum help regulate the scrotal temperature by lifting the testes up toward the body when it is cold and keeping the testes away from the body when warm. The walls of the scrotum have a number of layers, surrounded by wrinkled and pigmented skin. These include the dartos fascia, a layer of connective tissue containing smooth muscle fibers, and three layers of fascia derived from the muscular layers of the abdominal wall. The tunica vaginalis is a closed sac of membrane that contains a small amount of fluid to lubricate movement of the testis.

Body system:	reproductive system
Location:	at the base of the pelvis, behind and below the penis
Function:	maintains a suitable temperature to allow sperm production
Components:	skin, muscle, connective tissue, membrane
Related parts:	testes, penis

Blood Supply of the Testes

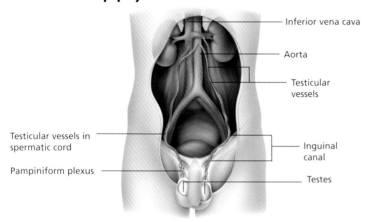

Inferior vena cava

Aorta

Testicular vessels

Testicular vessels in spermatic cord

Pampiniform plexus

Inguinal canal

Testes

During embryonic life, the testes develop within the abdomen; it is only after birth that they descend into their final position within the scrotum, and because of this, the blood supply of the testes arises from the abdominal aorta. The two long testicular arteries pass down the posterior abdominal wall until they enter the inguinal canal. As part of the spermatic cord they enter the scrotum where they supply the testes. They also form interconnections with the arteries to the vasa deferentia. Testicular veins arise from the testes and epididymes; instead of a single vein there is a network of veins, known as the pampiniform plexus, which acts as a heat exchange mechanism, cooling arterial blood before it reaches the testes.

Body system:	cardiovascular system
Location:	extends from the abdominal aorta down to the scrotum
Function:	provides blood rich in oxygen and nutrients to the tissues of the testes, and returns deoxygenated blood to the heart
Components:	testicular artery and vein, pampiniform plexus
Related parts:	aorta, inferior vena cava

The Penis

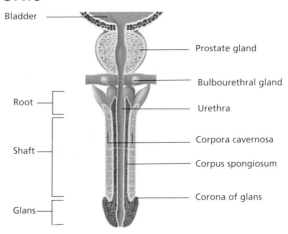

Bladder

Prostate gland

Bulbourethral gland

Root

Urethra

Shaft

Corpora cavernosa

Corpus spongiosum

Corona of glans

Glans

The penis is mainly composed of three columns of spongelike erectile tissue, the two corpora cavernosa and the corpus spongiosum. These are able to fill and become engorged with blood, causing an erection and enabling sexual intercourse to take place. The penis can be divided into three main sections: the root, which is made up of the expanded bases of the columns of erectile tissue covered by muscle fibers; the shaft, which is the main bulk of the penis and is formed of erectile tissue, connective tissue and blood and lymphatic vessels; and the glans, at the tip of which is the urethral orifice, the outlet for sperm and urine. The whole penis is covered with skin, which extends as a double layer (the foreskin) over the glans.

Body system:	reproductive system
Location:	at the base of the pubic bone
Function:	enables sexual intercourse to take place so that sperm can enter the female vagina; allows urination
Components:	root, shaft, glans, corpora cavernosa, corpus spongiosum
Related parts:	scrotum and testes

Cross-section of the Penis

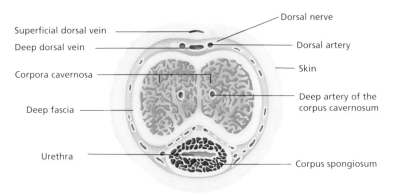

Superficial dorsal vein

Deep dorsal vein

Corpora cavernosa

Deep fascia

Urethra

Dorsal nerve

Dorsal artery

Skin

Deep artery of the corpus cavernosum

Corpus spongiosum

In a cross-section of the shaft of the penis, the relationship of the erectile tissue, blood vessels, and fascia can be seen more easily. The main bulk is made up of three masses of erectile tissue, the smaller corpus spongiosum and the two large corpora cavernosa. The corpus spongiosum contains within it the urethra, which is the tube that transports urine from the bladder and sperm from the testes. Each corpus cavernosum carries a deep central artery, which supplies the blood needed for an erection. A sleeve of connective tissue, the deep fascia, encloses the erectile tissue and the deep dorsal arteries, veins, and nerves. Outside the deep fascia is a layer of loose connective tissue that contains the superficial veins.

Body system:	reproductive system
Location:	at the base of the pubic bone
Function:	enables sexual intercourse to take place so that sperm can enter the female vagina, allows urination
Components:	corpora cavernosa, corpus spongiosum, dorsal veins arteries, nerves, connective tissue, skin, urethra
Related parts:	scrotum and testes

Superficial Perineal Muscles

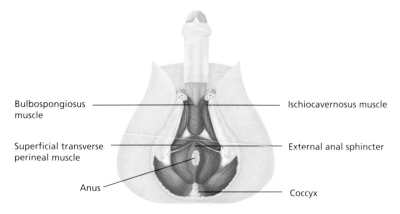

Bulbospongiosus muscle

Ischiocavernosus muscle

Superficial transverse perineal muscle

External anal sphincter

Anus

Coccyx

Three muscles are associated with the penis, although their fibers are confined to the root and structures around the penis rather than to the shaft or glans. These muscles are collectively known as the superficial perineal muscles because they lie in the perineum, the area around the anus and external genitalia. The bulbospongiosus muscle, which encircles the root of the penis, acts to compress the base of the corpus spongiosum, to help the urethra expel its contents. Contraction of the ischiocavernosus muscles at the bases of the corpora cavernosa, and the narrow superficial transverse perineal muscles that lie just under the skin in front of the anus, helps maintain erection of the penis.

Body system:	musculoskeletal system
Location:	in the perineum, the area surrounding the anus and the external genitalia
Function:	help maintain penile erection and expel contents of urethra
Components:	bulbospongiosus, ischiocavernosus, superficial transverse perineal muscles
Related parts:	pelvic floor muscles, bones of the pelvis

Blood Supply to the Penis

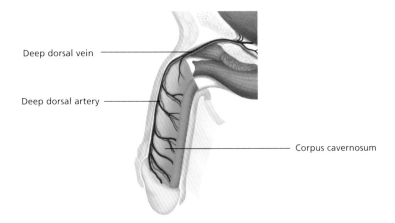

Deep dorsal vein

Deep dorsal artery

Corpus cavernosum

The arterial blood supply of the penis has two functions. As with any organ, it has to provide oxygenated blood for the tissues of the penis. However, it must also provide an additional supply to allow engorgement of the spongy erectile tissues. All the arteries supplying the penis originate from the internal pudendal arteries of the pelvis. The dorsal arteries lie on each side of the deep dorsal vein and supply connective tissue and skin. The deep arteries run within the corpora cavernosa to supply tissue there and to allow flooding of that tissue during erection. The deep dorsal vein of the penis receives blood from the cavernous spaces, while blood from the overlying connective tissue and skin is drained by the superficial dorsal veins.

Body system:	cardiovascular system
Location:	within the penis
Function:	provides blood rich in oxygen and nutrients to the tissues of the penis, and returns deoxygenated blood to the heart; floods the spongy tissue in the penis to cause erection
Components:	deep and superficial dorsal arteries and veins
Related parts:	internal pudendal arteries and veins

The Female Reproductive System

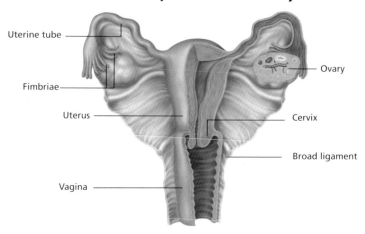

Uterine tube

Fimbriae

Uterus

Vagina

Ovary

Cervix

Broad ligament

The female reproductive tract consists of the internal genitalia—the ovaries, uterine (Fallopian) tubes, uterus, cervix, and vagina—and the external genitalia, or vulva. The almond-shaped ovaries lie on either side of the uterus suspended by ligaments. They store the female eggs, or ova, and release one a month into the uterine tube, which connects the ovary to the uterus. The uterus is a muscular, pear-shaped, hollow organ, whose function is to nurture and protect the growing fetus. Between the uterus and the vagina is the cervix, which has a narrow central canal. The upper surface of the uterus and ovaries is draped in a "tent" of peritoneum, (the lining of the abdominal cavity), which helps keep the uterus in position.

Body system:	reproductive system
Location:	within the pelvis
Function:	stores and releases eggs (ova) for fertilization, nurtures and protects the growing fetus, enables childbirth to occur, produces hormones
Components:	ovaries, uterine tubes, uterus, cervix, vagina, vulva
Related parts:	pelvis, bladder

Position in the Pelvis

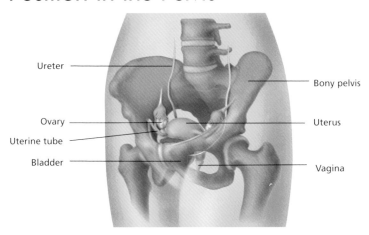

Ureter

Bony pelvis

Ovary

Uterus

Uterine tube

Bladder

Vagina

In adult women, the internal genitalia (which, apart from the ovaries, are basically tubular in structure) are located deep within the pelvic cavity. They are thus protected by the presence of the circle of bone that makes up the pelvis—the hip bones, the sacrum, and the coccyx. This is in contrast to the pelvic cavity of children, which is relatively shallow. The adult uterus lies between the bladder and the rectum, although its position does change with movement. Normally, the uterus tilts forward on top of the bladder, a position known as anteversion. In some women, however, the uterus curves forward more (anteflexion) or bends not forward but backward toward the rectum (retroflexion).

Body system:	reproductive system
Location:	within the pelvis
Function:	stores and releases eggs (ova) for fertilization, nurtures and protects the growing fetus, enables childbirth to occur, produces hormones
Components:	ovaries, uterine tubes, uterus, cervix, vagina, vulva
Related parts:	pelvis, bladder, rectum

Blood Supply to the Genitalia

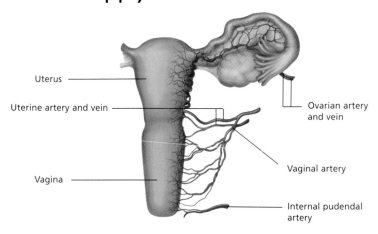

Uterus

Uterine artery and vein

Ovarian artery and vein

Vaginal artery

Vagina

Internal pudendal artery

The female reproductive organs receive a rich blood supply via an interconnecting network of arteries. The ovarian artery originates at the abdominal aorta and passes through the mesovarium, the fold of peritoneum in which the ovary lies, to supply the ovary and uterine tube. The uterine artery is a branch of the large internal iliac artery of the pelvis and approaches the uterus at the level of the cervix. Together with the uterine artery, the vaginal artery supplies blood to the vaginal walls, while the lower third of the vagina and the anus are supplied by the internal pudendal artery. A network of veins in the walls of the uterus and vagina drain blood back into the internal iliac vein via the uterine vein.

Body system :	cardiovascular system
Location:	surrounding, and within the walls of, the organs of the female reproductive system
Function:	supplies blood rich in nutrients and oxygen to the tissues and takes deoxygenated blood back to the heart
Components:	ovarian artery and vein, vaginal artery, internal pudendal artery, uterine artery and vein
Related parts:	internal iliac artery and vein

Female External Genitalia

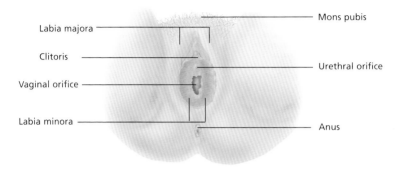

Mons pubis

Labia majora

Clitoris

Vaginal orifice

Labia minora

Urethral orifice

Anus

The female external genitalia, or vulva, are those parts that are external to the vagina. The mons pubis is the rounded, fatty area that lies above the pubic bone. Following puberty, this area is usually covered with coarse pubic hair. The two outer folds of skin that lie across, and protect, the vulval opening are called the labia majora (large lips), while the smaller more delicate folds inside the cleft of the vulva are known as the labia minora (small lips). Inside the labia minora is an area called the vestibule, which contains the urethral orifice and the vaginal opening. At the top of the vestibule is the clitoris, a structure that is formed mainly of erectile tissue and which is analogous to the penis in males.

Body system:	reproductive system
Location:	external to the vagina
Function:	vaginal orifice is the passageway between the internal and external genitalia; labia protect and cover vestibule.
Components:	mons pubis, labia majora and minora, clitoris, vaginal orifice, vestibule
Related parts:	internal genitalia, anus

The Uterus

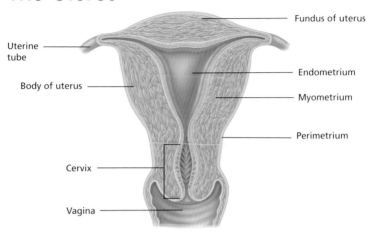

Fundus of uterus

Uterine tube

Endometrium

Body of uterus

Myometrium

Perimetrium

Cervix

Vagina

In in the nonpregnant state, the uterus is about 3 in (7.5 cm) long and 2 in (5 cm) across at its widest point, but expands to accommodate a growing fetus. The body of the uterus forms its main bulk and has a central triangular space from which extend the uterine (Fallopian) tubes. The cervix is the lower part of the uterus that projects into the vagina. The thick wall is composed of three layers: the perimetrium is the thin outer coat, which is continuous with the pelvic peritoneum; the myometrium is the muscular middle layer and contains most of the blood vessels and nerves that supply the uterus; the endometrium is the lining of the uterus, which becomes thicker during the menstrual cycle in preparation for an embryo.

Body system:	reproductive system
Location:	within the pelvis, behind the bladder
Function:	nurtures and protects the growing fetus
Components:	perimetrium, myometrium, endometrium
Related parts:	uterine tubes, vagina

The Uterus in Pregnancy

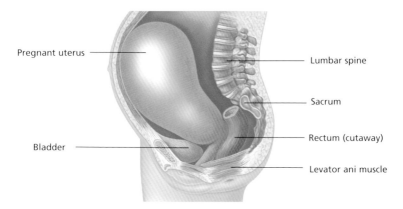

Pregnant uterus

Lumbar spine

Sacrum

Rectum (cutaway)

Bladder

Levator ani muscle

During pregnancy, the uterus increases in size and fills much of the space in the abdominal cavity. Pressure of the enlarged uterus pushes the abdominal organs up against the diaphragm, encroaching on the thoracic cavity. Organs such as the stomach and bladder are compressed to such an extent in late pregnancy that their capacity is diminished. Hence, a woman will feel "full" much sooner when eating and will pass urine more often. After pregnancy, the uterus rapidly decreases in size again, but it never returns to its prepregnant state. In the final stages of pregnancy the uterus will have increased in weight from a prepregnant 16 oz (453 g) to around 32 oz (900 g) as the muscle fibers in the myometrium increase in size and number.

Body system:	reproductive system
Location:	within the pelvis
Function:	nurtures and protects the growing fetus
Components:	perimetrium, myometrium, endometrium
Related parts:	uterine tubes, vagina, abdominal organs

The Vagina

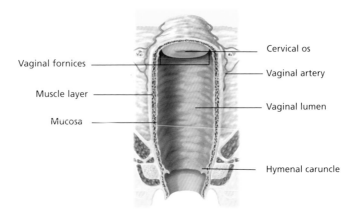

Vaginal fornices

Muscle layer

Mucosa

Cervical os

Vaginal artery

Vaginal lumen

Hymenal caruncle

The vagina is a thin-walled muscular tube that extends from the cervix to the external genitalia. It forms the main part of the birth canal and receives the penis during sexual intercourse. The front and back walls of the vagina normally lie in contact with one another, closing the lumen (central space), although the vagina can expand greatly, for example, during childbirth. The cervix, the lower end of the uterus, projects down into the lumen of the vagina, and at the junction of the two structures are recesses known as the vaginal fornices. The wall of the vagina has three layers: the outer layer of fibroelastic connective tissue; the central muscular layer; and the inner mucosal layer, which is thrown into deep folds called rugae.

Body system :	reproductive system
Location:	extends from the cervix of the uterus to the external genitalia
Function:	receives the penis during sexual intercourse, forms the main part of the birth canal
Components:	outer layer of connective tissue, middle muscle layer, inner mucosal layer
Related parts:	cervix, external genitalia

The Cervix

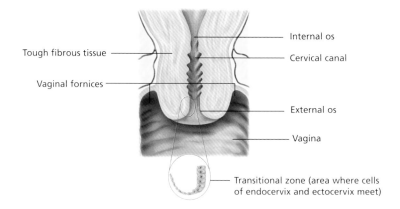

Internal os

Cervical canal

Tough fibrous tissue

Vaginal fornices

External os

Vagina

Transitional zone (area where cells of endocervix and ectocervix meet)

The cervix is the narrowed lower part of the uterus that projects down into the vagina. Running down the center of the cervix is a narrow canal, approximately 1 in (2.5 cm) long in adult women, which is the downward continuation of the uterine cavity, and opens into the vagina at its lower end, the external os. The walls of the cervix are tough and contain much fibrous tissue, as well as muscle, unlike the body of the uterus which is mainly muscle. The epithelium, or lining, of the cervix is of two types. The endocervix is the lining of the cervical canal and contains mucus-secreting glands. The portion of the cervix that projects down into the vagina is composed of squamous epithelium and has many layers.

Body system:	reproductive system
Location:	the lower end of the uterus, projecting into the vagina
Function:	anchors the uterus, provides a communication channel between the uterus and vagina
Components:	fibrous tissue, central canal containing folds of mucosa
Related parts:	body of uterus, vagina

The Ovaries

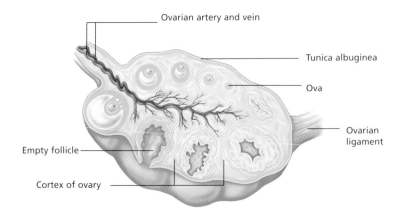

Ovarian artery and vein

Tunica albuginea

Ova

Ovarian ligament

Empty follicle

Cortex of ovary

The paired ovaries are situated in the lower abdomen and lie on either side of the uterus, anchored to it by the ovarian ligaments. At birth, a girl's ovaries contain a lifetime's supply of immature ova, or eggs, which can be fertilized by sperm to produce embryos. The ovaries also produce the hormones necessary for female development. Each ovary is surrounded by a protective layer of fibrous tissue, the tunica albuginea, and has a central region containing blood vessels and nerves (the medulla) and an outer cortex within which the ova develop. This cross-section shows the follicles containing ovum at different stages of development. During ovulation, only one follicle will mature to release an ovum into the uterine tube.

Body system:	reproductive system
Location:	either side of the uterus
Function:	produce female sex hormones, store ova, and release them into uterine tube
Components:	tunica albuginea, cortex, medulla, follicles containing ova
Related parts:	uterine tubes, uterus, uterine ligaments

Supporting Ligaments

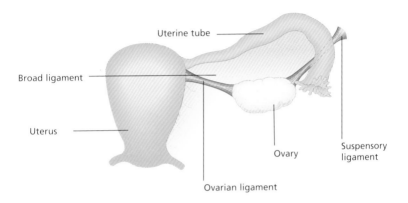

Uterine tube

Broad ligament

Uterus

Ovary

Suspensory ligament

Ovarian ligament

Each ovary is held in position in the pelvis by several ligaments. The broad ligament is a tentlike fold of pelvic peritoneum that hangs down on either side of the uterus, enclosing the uterine tubes and ovaries. The ovary is anchored to the side wall of the pelvis by a section of the broad ligament known as the suspensory ligament, which also carries the ovarian blood and lymphatic vessels. The ovarian ligament is a fibrous cord that runs within the broad ligament from the ovary to the side of the uterus, just below the point of entry of the uterine tube. Any of these ligaments may become stretched during pregnancy and childbirth, which means that the position of the ovary may change from its prepregnancy position.

Body system:	reproductive system
Location:	surrounding the ovary, uterine tube, and uterus
Function:	hold the ovaries in position within the pelvis
Components:	broad ligament, ovarian ligament, suspensory ligament
Related parts:	ovaries, uterine tubes, uterus

The Uterine Tubes

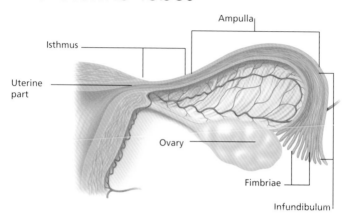

Isthmus

Ampulla

Uterine part

Ovary

Fimbriae

Infundibulum

The uterine, or Fallopian, tubes collect the ova (eggs) released from the ovaries and transport them to the uterus. They also provide a site for fertilization of an ovum by a sperm. Each uterine tube is about 4 in (10 cm) long and extends outward from the upper part of the uterus to the wall of the pelvic cavity. Anatomically, the tube is divided into four parts. The infundibulum is the funnel-shaped outer end of the tube, which has fingerlike projections called fimbriae that hang over the ovary, ready to scoop up an ovum at ovulation. The ampulla is the longest and widest part and is the usual site of fertilization. Close to the uterus is the isthmus, a thick-walled segment that leads into the shortest section of the tube, the uterine part.

Body system:	reproductive system
Location:	extends from the uterus laterally toward the pelvic wall
Function:	transport ova to the uterus following ovulation; provide site for fertilization
Components:	infundibulum, ampulla, isthmus, uterine part
Related parts:	ovaries, uterus, ligaments

The Placenta

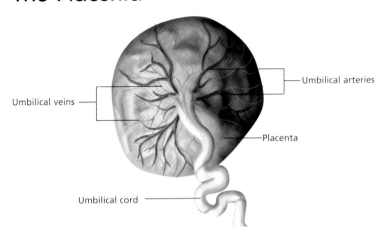

Umbilical veins

Umbilical arteries

Placenta

Umbilical cord

During pregnancy, the placenta takes on the role of the lungs and the intestine for the developing fetus. It achieves this by bringing the blood of the fetus close to the maternal blood within its internal structure, allowing the fetus to take up oxygen and nutrients, while waste products are removed. At full term, the placenta is a deep red, round or oval, flattened organ. It weighs about 18 oz (510 g), or one-sixth of the weight of the fetus it nourishes. There are two sides to the placenta—the maternal aspect and the fetal aspect. Throughout pregnancy, the maternal aspect of the placenta is firmly attached to the uterine lining. The fetal aspect, from which the umbilical cord arises, is covered in fetal membranes and large blood vessels.

Body system:	reproductive system
Location:	attached directly to the uterine lining, and to the developing fetus by the umbilical cord
Function:	provides nutrients and oxygen to the fetus, and removes waste products
Components:	maternal and fetal blood vessels, connective tissue, membrane
Related parts:	uterus

249

Inside the Placenta

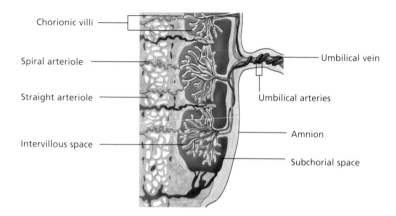

Chorionic villi

Spiral arteriole

Straight arteriole

Intervillous space

Umbilical vein

Umbilical arteries

Amnion

Subchorial space

A cross-section of the placenta reveals that the organ is made up partly from maternal tissue and partly from fetal tissue. The spiral arteries that arise from the maternal uterine arteries bring blood into the placenta. This blood then fills wide "pools" (intervillous spaces) in which the fetal villi are suspended. The fetal chorionic villi are fingerlike projections that contain blood vessels connected to the fetus through the umbilical cord. They branch repeatedly to create the maximum amount of surface area for the transfer of oxygen, nutrients, and waste substances. Although the two circulations come close to each other, maternal and fetal blood do not mix and are separated by the thin walls of the villi.

Body system:	reproductive system
Location:	attached to the uterus during pregnancy
Function:	supplies fetus with oxygen, disposes of fetal waste products, produces hormones to maintain pregnancy and prepare for birth
Components:	maternal and fetal blood vessels, intervillous spaces, chorionic villi
Related parts:	uterus

The Gluteus Maximus

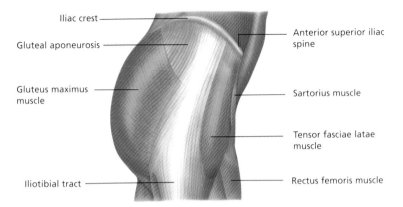

Iliac crest

Gluteal aponeurosis

Gluteus maximus muscle

Iliotibial tract

Anterior superior iliac spine

Sartorius muscle

Tensor fasciae latae muscle

Rectus femoris muscle

The gluteal region, or buttock, lies behind the pelvis. Its shape is formed by a number of large muscles that help stabilize and move the hip joint, all covered by a layer of fat. The gluteus maximus is one of the largest muscles in the body and covers the other gluteal muscles, with the exception of about one-third of the smaller gluteus medius. It arises from the ileum (part of the pelvis) and the back of the sacrum and its fibers run down and outward at a 45° angle toward the femur. The main function of the gluteus maximus is to straighten the legs, as when standing up from a sitting position. When standing, the muscle covers the bony ischial tuberosity; when sitting, the muscle moves up and away from the tuberosity.

Body system:	musculoskeletal system
Location:	behind the pelvis, in the gluteal, or buttock, region
Function:	extends (straightens) the leg, as when standing up from a sitting position
Components:	thick, coarse muscle fibers
Related parts:	bones of pelvis, other gluteal muscles

Deep Gluteal Muscles

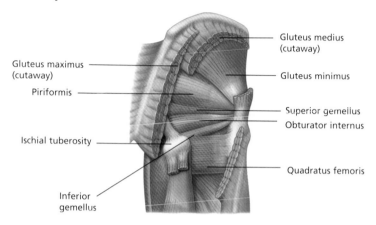

Gluteus medius (cutaway)

Gluteus maximus (cutaway)

Piriformis

Gluteus minimus

Superior gemellus

Obturator internus

Ischial tuberosity

Quadratus femoris

Inferior gemellus

Beneath the gluteus maximus lie a number of other muscles that act to stabilize the hip joint and move the leg. The gluteus medius and maximus are both fan-shaped muscles with fibers that run in the same direction. Together these muscles have an essential role in the action of walking and hold the pelvis level when one foot is lifted off the ground, rather than let it sag to that side. This allows the non-weightbearing foot to clear the ground before being swung further forward. Other muscles in this area act mainly to help rotate the lower limb laterally and stabilize the hip joint. These muscles include the piriformis, the obturator internus, the superior and inferior gemelli, and the quadratus femoris.

Body system :	musculoskeletal system
Location:	deep to the gluteus maximus in the buttock region
Function:	act to stabilize the hip and pelvis, and play a role in walking and rotating the thigh laterally
Components:	gluteus medius, gluteus minimus, piriformis, superior and inferior gemelli, quadratus femoris
Related parts:	gluteus maximus, bones of pelvis

Gluteal Bursae

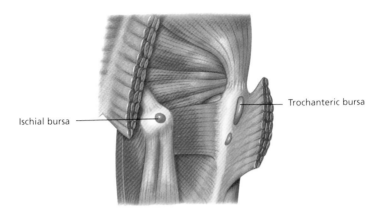

Trochanteric bursa

Ischial bursa

A bursa is a small, fluid-filled sac, similar to an underfilled water bottle. Bursae are found between two structures, usually bone and tendon, that regularly move against each other, protecting them from wear and tear. There are three main groups of bursae in the gluteal region. The large trochanteric bursae lie between the thick, upper fibers of the gluteus maximus muscle and the greater trochanter of the upper thigh bone (femur). The ischial bursa lies between the lower fibers of the gluteus maximus muscle and the ischial tuberosity, the part of the pelvis that bears the body's weight when sitting. The gluteofemoral bursa (not shown) lies on the outer side of the leg, between the gluteus maximus and vastus lateralis muscles.

Body system:	musculoskeletal system
Location:	between the tendons and bones in the gluteal (buttock) region
Function:	provide a "cushion" where two structures regularly rub against each other
Components:	trochanteric, ischial, gluteofemoral bursae
Related parts:	gluteal muscles, pelvic bones

The Hip Joint

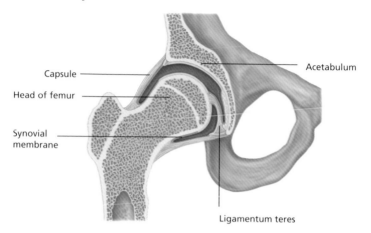

Capsule

Head of femur

Synovial membrane

Acetabulum

Ligamentum teres

The hip joint is the strong ball-and-socket joint that connects the lower limb to the pelvis. The head of the femur (thigh bone) is the "ball" that fits tightly into the "socket" formed by the deep, cuplike acetabulum of the hip bone of the pelvis. The articular surfaces—the parts of the bone that come into contact with each other—are covered with a protective layer of hyaline cartilage, which is very slippery. The hip joint is a synovial joint, which means that movement is further lubricated by a thin layer of synovial fluid. This fluid lies between the articular surfaces within the synovial cavity and is secreted by the synovial membrane that lines the entire hip joint and its surrounding capsule.

Body system:	musculoskeletal system
Location:	connects the lower limb to the pelvis
Function:	enables a wide range of movement
Components:	head of femur, acetabulum of hip bone, capsule, synovial membrane, and fluid, hyaline cartilage
Related parts:	lower limbs, pelvis, spine

Ligaments of the Hip Joint

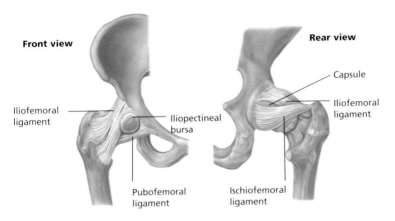

Front view

Rear view

Iliofemoral
ligament

Iliopectineal
bursa

Pubofemoral
ligament

Capsule

Iliofemoral
ligament

Ischiofemoral
ligament

The hip joint is enclosed and protected by a thick, fibrous capsule, which is strengthened and stabilized by a number of tough ligaments. These ligaments are thickened parts of the joint capsule, which extends from the rim of the acetabulum to the neck of the femur. They generally follow a spiral path from the hip bone to the femur and are named according to the bone to which they attach. The iliofemoral ligament is a strong Y-shaped structure that supports the front of the hip joint and prevents it from overextending, as does the large, spiral ischiofemoral ligament that lies at the back of the joint. The triangular pubofemoral ligament is located at the front of the hip joint and prevents overabduction.

Body system:	musculoskeletal system
Location:	extend from the hip bone to the femur
Function:	strengthen and stabilize the capsule of the hip joint
Components:	iliofemoral, ischiofemoral, pubofemoral ligaments
Related parts:	hip bone, hip joint, femur

The Femur

Rear view (left leg)

Front view (left leg)

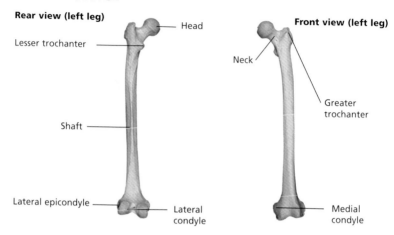

Head

Lesser trochanter

Neck

Shaft

Greater trochanter

Lateral epicondyle

Lateral condyle

Medial condyle

The longest and heaviest bone in the body is the femur, or thigh bone. Approximately 18 in (45 cm) in length in adult males, the femur forms about one-quarter of a person's total height. It has a long, thick shaft with two expanded ends, the upper of which slots neatly into the acetabulum of the hip bone to form the hip joint. The lower end articulates with the tibia and fibula to form the knee joint. The upper end of the femur is anatomically divided into the head, neck, and greater and lesser trochanters—projections of bone that allow the attachment of muscles. The lower end is made up of two enlarged bony processes, the medial and lateral condyles, which articulate with the two bones of the lower leg.

Body system:	musculoskeletal system
Location:	extends from the hip bone in the pelvis to the knee joint
Function:	supports and stabilizes the lower body, enabling walking and other movements
Components:	head, neck, greater and lesser trochanters, shaft, condyles, and epicondyles
Related parts:	hip bone, other bones of pelvis, tibia, and fibula

Internal Structure of the Femur

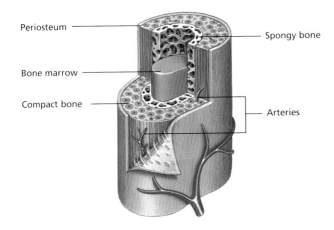

Periosteum

Spongy bone

Bone marrow

Compact bone

Arteries

The femur is one of the bones in the body that are classified as long bones. Bones of this type have a relatively long shaft, or diaphysis, and two expanded ends, or epiphyses. The bone is entirely covered with a tough, protective membrane, the periosteum, which is nourished by tiny arteries in the bone tissue. The diaphysis of the femur is a tube composed of compact bone, which is strong and dense. This layer of compact bone encloses a core of yellow bone marrow that, in adults, consists of fat cells. The expanded ends of the femur are formed of a surface layer of compact bone that surrounds a central area of spongy bone. This central area is much looser in structure and there is no marrow in the epiphyses.

Body system:	musculoskeletal system
Location:	extends from the hip bone in the pelvis to the knee joint
Function:	supports and stabilizes the lower body, enabling walking and other movements
Components:	periosteum, compact bone, spongy bone, bone marrow
Related parts:	hip bone, other bones of pelvis, tibia, and fibula

Muscle Attachments of the Femur

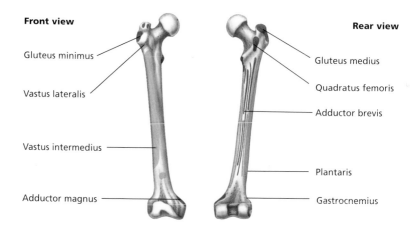

Front view

Gluteus minimus

Vastus lateralis

Vastus intermedius

Adductor magnus

Rear view

Gluteus medius

Quadratus femoris

Adductor brevis

Plantaris

Gastrocnemius

The femur (thigh bone) is a very strong bone that provides sites of attachment for many of the muscles of locomotion in the hip joint and leg. Some muscles, such as the powerful gluteus muscles, have their origins on the pelvic bones and so cross the hip joint to insert into the femur. When these muscles contract, they cause the hip joint to move, allowing the leg to bend, straighten, or move sideways. Other muscles originate on the femur itself and pass down across the knee joint to insert into the tibia or fibula. These muscles allow the knee to bend or straighten. Where muscle is attached to bone, there s a visible projection, or bony process. The surface of the bone in these areas can also become roughened.

Body system:	musculoskeletal system
Location:	connects the lower limb to the pelvis
Function:	enables a wide range of movement
Components:	wide variety of attachments for leg muscles (some shown above)
Related parts:	muscles of the lower limb, pelvis, tibia, and fibula

The Tibia and the Fibula

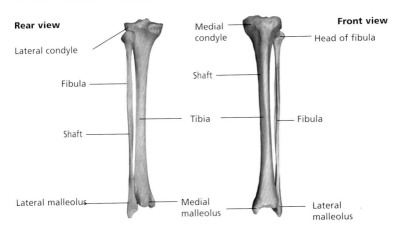

Rear view

Lateral condyle

Fibula

Shaft

Lateral malleolus

Medial condyle

Shaft

Tibia

Medial malleolus

Front view

Head of fibula

Fibula

Lateral malleolus

Second only in size to the femur (thigh bone), the tibia (shin bone) has the shape of a typical long bone with an elongated shaft and two expanded ends. The tibia lies alongside the fibula, a long narrow bone that has none of the strength of the tibia. Whereas the upper end of the tibia is expanded to form the lateral and medial condyles, which form part of the knee joint with the femur, the fibula articulates only with the tibia at its upper end. It is, however, an important support for the ankle. At the lower end of the fibula is the lateral malleolus, a protuberance that articulates with the talus (ankle bone). The medial malleolus of the tibia also connects with the talus and bears four-fifths of the forces transmitted from the foot.

Body system:	musculoskeletal system
Location:	extend from the knee to the ankle
Function:	strengthen and stabilize the lower limb, enable movement
Components:	shaft, lateral and medial condyles, lateral and medial malleoli
Related parts:	femur, talus

Cross-section of the Tibia and Fibula

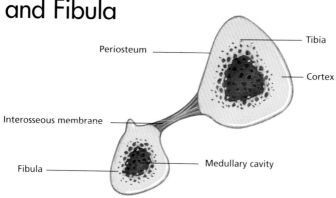

Periosteum

Tibia

Cortex

Interosseous membrane

Fibula

Medullary cavity

The shafts of the tibia and fibula are roughly triangular in cross-section. The tibial shaft is much greater in diameter than that of the fibula as it is the main weight-bearing bone of the lower leg. The fibula acts as a strut, increasing the stability of the lower leg under load. The tibia and fibula have a typical long bone structure with a thick, tubular outer cortex surrounding a spongy medullary cavity. Their hollow structure provides maximal mechanical strength with minimal support material, namely dense cortical bone. The two bones are enveloped in a tough membranous layer called periosteum. The periosteum from the borders of the two bones blends to form the interosseous membrane, which anchors them together.

Body system:	musculoskeletal system
Location:	extend from the knee to the ankle
Function:	supports and stabilizes the lower limb, enabling movement
Components:	head, shaft, medial and lateral condyles, medial and lateral malleolus, interosseous membrane
Related parts:	femur, knee joint, talus

Ligaments of the Tibia and Fibula

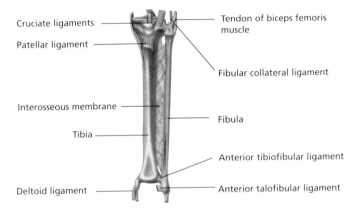

Cruciate ligaments

Patellar ligament

Interosseous membrane

Tibia

Deltoid ligament

Tendon of biceps femoris muscle

Fibular collateral ligament

Fibula

Anterior tibiofibular ligament

Anterior talofibular ligament

A number of strong fibrous bands called ligaments surround the tibia and fibula and bind them to each other and to other bones in the leg. Just under the knee is the upper joint between the head of the fibula and the underside of the lateral tibial condyle. This joint is surrounded and protected by a fibrous joint capsule, which is strengthened by the anterior and posterior tibiofibular ligaments. Other ligaments bind the bones of the lower leg to the femur. The strongest of these are the medial and lateral ligaments of the knee joint, which run vertically down from the femur to the corresponding bone beneath. The lower ends of the tibia and fibula are bound tightly together by ligaments to maintain the stability of the ankle joint.

Body system:	musculoskeletal system
Location:	around the knee and ankle joints, binding bones together
Function:	strengthen and stabilizes the bones and joints in the lower limb
Components:	cruciate ligaments, collateral ligaments, patellar ligaments
Related parts:	hip bone, other bones of pelvis, tibia and fibula

The Knee Joint and the Patella

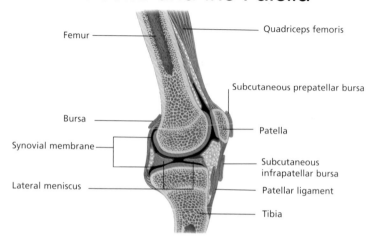

Femur

Quadriceps femoris

Subcutaneous prepatellar bursa

Bursa

Patella

Synovial membrane

Subcutaneous infrapatellar bursa

Lateral meniscus

Patellar ligament

Tibia

The knee is the joint between the lower end of the femur (thigh bone) and the upper end of the tibia. It is a synovial joint, in which movement is lubricated by fluid that is secreted by the synovial membrane lining the joint cavity. At the front of the joint is the patella, or kneecap, a flattened disk of bone that lies within the tendon of the quadriceps femoris thigh muscle and protects it from wear and tear. In front of the patella, and just below it, are bursae (fluid-filled sacs) that protect the patella when kneeling. Although the "fit" between the femoral condyles (bulbous ends of the bone) and the tibia is not a tight one, the knee is a reasonably stable joint, especially because it is surrounded by strong muscles and ligaments.

Body system:	musculoskeletal system
Location:	connects the femur to the lower leg
Function:	enables a wide range of movement in the lower limbs
Components:	femoral condyles, tibia, patella, bursae, synovial membranes, menisci
Related parts:	muscles of lower limbs

The Menisci

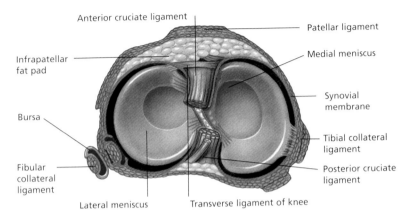

Anterior cruciate ligament

Patellar ligament

Infrapatellar
fat pad

Medial meniscus

Synovial
membrane

Bursa

Tibial collateral
ligament

Fibular
collateral
ligament

Posterior cruciate
ligament

Lateral meniscus

Transverse ligament of knee

Looking down on the upper surface of the tibia within the opened knee, the two C-shaped menisci can clearly be seen. They are plates of tough fibrocartilage that lie upon the articular surface of the tibia, deepening the depression into which the femoral condyles fit. The menisci also have the function of acting as shock absorbers within the knee and help to prevent the side-to-side rocking of the joint. The two menisci are wedge-shaped in cross-section, their external margins being widest. Centrally, they taper to a thin unattached edge. Anteriorly, the two menisci are attached to each other by the transverse ligament of the knee, while the outer edges of the menisci are firmly attached to the joint capsule.

Body system:	musculoskeletal system
Location:	lying on the articular surface of the tibia in the knee joint
Function:	deepen the depression into which the femoral condyles fit; act as shock absorbers
Components:	lateral and medial menisci
Related parts:	tibia, femur, ligaments

Extracapsular Knee Ligaments

Front view of flexed knee

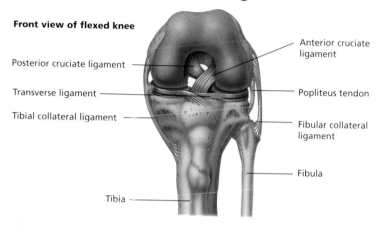

Anterior cruciate ligament

Posterior cruciate ligament

Transverse ligament

Tibial collateral ligament

Popliteus tendon

Fibular collateral ligament

Fibula

Tibia

Unlike the bones of the hip joint, the bones of the knee do not fit together in a particularly stable fashion. For this reason, the stability of the knee joint depends to a great extent upon the ligaments and muscles that surround it. The joint cavity of the knee is enclosed in a fibrous capsule; the ligaments that support the knee can be divided into two groups (extracapsular and intracapsular ligaments), depending on their relationship to this capsule. The extracapsular ligaments lie outside the capsule and act to prevent the lower leg bending forward at the knee (hyperextension). Between them, they bind the femur to the fibula and tibia, strengthen the front and the back of the knee and prevent abnormal movement.

Body system:	musculoskeletal system
Location:	surrounding the knee
Function:	support and stabilize knee, prevent abnormal movement
Components:	quadriceps tendon, fibular collateral, tibial collateral, oblique popliteal, arcuate popliteal, transverse ligaments
Related parts:	tibia, fibula, femur, meniscus

Intracapsular Knee Ligaments

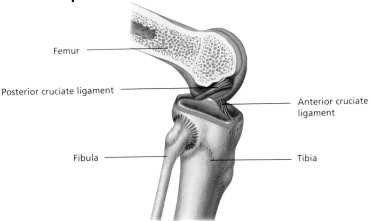

Femur

Posterior cruciate ligament

Anterior cruciate ligament

Fibula

Tibia

The intracapsular ligaments connect the tibia to the femur within the center of the knee joint, inside the capsule itself, and prevent forward and backward displacement of the tibia and femur. The two main intracapsular ligaments are known as the cruciate ligaments because they cross over each other and form the shape of a cross. The anterior cruciate is slack when the knee is flexed (bent) and taut when the knee is extended (straightened). The posterior cruciate ligament is the stronger of the two and tightens during flexion (bending), preventing overflexion of the knee joint. It is very important for the stability of the knee when bearing weight in a flexed position (for example, when walking downhill).

Body system:	musculoskeletal system
Location:	inside the capsule of the knee joint
Function:	prevent forward and backward displacement of the femur and tibia
Components:	anterior and posterior cruciate ligaments
Related parts:	femur, tibia

Bursae of the Knee

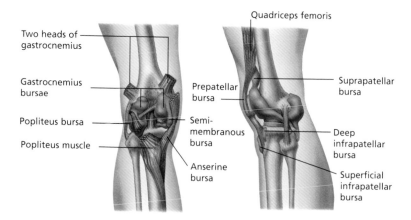

Two heads of gastrocnemius

Gastrocnemius bursae

Popliteus bursa

Popliteus muscle

Prepatellar bursa

Semi-membranous bursa

Anserine bursa

Quadriceps femoris

Suprapatellar bursa

Deep infrapatellar bursa

Superficial infrapatellar bursa

Bursae are small sacs filled with synovial fluid that reduce the friction between bones and tendons as they slide over one another. There are a number of bursae around the knee, some of which are continuous with the joint cavity—the fluid-filled space between the articular surfaces. The suprapatellar bursa lies above the joint cavity, between the lower end of the femur and the powerful quadriceps femoris muscle. The prepatellar and infrapatellar bursae surround the patella and the patellar ligament. The prepatellar bursa allows the skin to move freely over the patella during movement. The superficial and deep infrapatellar bursae lie around the lower end of the patellar ligament where it attaches to the tibial tuberosity.

Body system:	musculoskeletal system
Location:	around the knee joint
Function:	act as a cushion between tendons and bones, preventing wear and tear due to friction
Components:	popliteus, anserine, semimembranous, prepatellar, suprapatellar, infrapatellar bursae
Related parts:	femur, tibia, fibula, tendons, muscles

Anterior Thigh Muscles

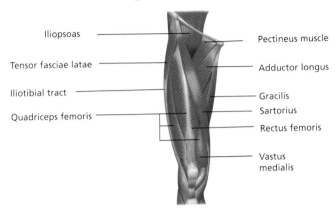

Iliopsoas

Tensor fasciae latae

Iliotibial tract

Quadriceps femoris

Pectineus muscle

Adductor longus

Gracilis

Sartorius

Rectus femoris

Vastus medialis

The muscles of the anterior compartment of the thigh flex (bend) the hip and extend (straighten) the knee, the actions associated with walking. The large iliopsoas muscle is a powerful muscle that flexes the thigh, bringing the knee up and forward. It arises partly from the inside of the pelvis and partly from the lower vertebrae and inserts into the projection of the upper femur known as the lesser trochanter. The tensor fasciae latae muscle inserts into the strong band of connective tissue that runs down the outside of the leg and helps to support the femur on the tibia during standing. The sartorius is the longest muscle in the body and runs as a flat strap across the thigh from the pelvis to the top of the tibia.

Body system:	musculoskeletal system
Location:	anterior (front) compartment of the thigh
Function:	flex the hip, extend the knee
Components:	iliopsoas, tensor fasciae latae, quadriceps femoris, sartorius, gracilis, pectineus muscles
Related parts:	pelvis, femur, tibia

The Quadriceps Femoris

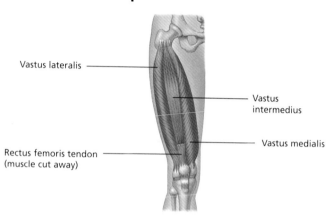

Vastus lateralis

Vastus intermedius

Vastus medialis

Rectus femoris tendon
(muscle cut away)

The quadriceps femoris is the large four-headed muscle that makes up the bulk of the thigh and acts to straighten the knee. It is one of the most powerful muscles in the body and consists of four muscles whose tendons combine to form the strong quadriceps tendon. This tendon inserts into the top of the patella and continues down as the patellar tendon to the top of the tibia. The rectus femoris is a straight muscle that overlies the other three muscles and helps flex the hip joint and straighten the knee. The vastus lateralis is the largest component of the quadriceps muscle and lies on the outer thigh, while on the inner side is the vastus medialis. The vastus intermedius lies centrally underneath the rectus femoris.

Body system :	musculoskeletal system
Location:	extends from the head of the femur to the knee
Function:	straightens the knee
Components:	vastus lateralis, rectus femoris, vastus intermedius, vastus medialis
Related parts:	femur, tibia

Hamstrings

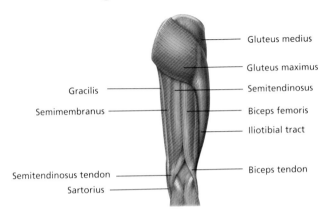

Gluteus medius

Gluteus maximus

Semitendinosus

Gracilis

Biceps femoris

Semimembranus

Iliotibial tract

Biceps tendon

Semitendinosus tendon

Sartorius

The three large muscles of the posterior thigh (commonly known as the hamstrings) extend the hip and flex the knee, although they cannot do both at the same time. The biceps femoris has two heads: the longer head arises from the ischial tuberosity of the pelvis and the shorter head arises from the back of the femur. The semitendinosus also arises in the pelvis and is named for its unusually long tendon, which begins about two-thirds of the way down its course to the inner side of the upper tibia. The semimembranus muscle also arises from the ischial tuberosity and runs down the back of the thigh, deep to the semitendinosus, to insert into the inner side of the upper tibia.

Body system:	musculoskeletal system
Location:	posterior compartment of thigh
Function:	extend the hip and flex the knee
Components:	biceps femoris, semitendinosus, semimembranus
Related parts:	pelvis, femur, tibia

The Adductor Muscles

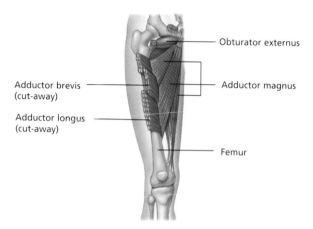

Obturator externus

Adductor brevis
(cut-away)

Adductor magnus

Adductor longus
(cut-away)

Femur

The muscles of the inner thigh are known as adductors because they enable adduction of the leg (moving the lower limb in toward the midline). These muscles arise from the lower part of the pelvis and insert into the femur at various levels. The adductor longus is a large fan-shaped muscle that lies in front of the other adductors and has a palpable tendon in the groin. The adductor brevis is a shorter muscle that lies under the adductor longis. The adductor magnus is a large triangular muscle that acts as both an adductor muscle and a hamstring. The gracilis (not shown) is a straplike muscle that runs vertically down the inner thigh. Deep within this group of adductors is the small obturator externus muscle.

Body system:	musculoskeletal system
Location:	inner thigh, extending from the pelvis to the femur
Function:	move the leg back toward the midline
Components:	adductor magnus, adductor brevis, adductor longus, obturator externus
Related parts:	pelvis, femur

Anterior Muscles of the Lower Leg

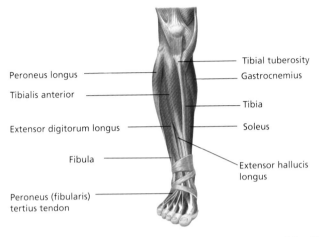

Peroneus longus

Tibialis anterior

Extensor digitorum longus

Fibula

Peroneus (fibularis)
tertius tendon

Tibial tuberosity

Gastrocnemius

Tibia

Soleus

Extensor hallucis
longus

The anterior group of muscles of the lower leg lie in front of the tibia. They all have a similar action in that they are dorsiflexors of the foot. This means that when they contract, they bring the toes up and the heel down. The tibialis anterior muscle can be felt under the skin alongside the tibia, and its tendon is easily seen in the ankle region. Under the tibialis anterior lies the extensor digitorum longus, which attaches to the outer four toes. The peroneus (fibularis) tertius is not always present but when it is, it may join the extensor digitorum longus muscle. It inserts into the fifth metatarsal bone near the little toe. The thin extensor hallucis longus muscle runs down to insert into the end of the hallux (big toe).

Body system:	musculoskeletal system
Location:	in front of the tibia in the lower leg
Function:	dorsiflex the foot (bring the toes up and the heel down)
Components:	tibialis anterior, extensor digitorum longus, peroneus (fibularis) tertius, extensor hallucis longus
Related parts:	tibia, fibula, bones of foot

271

Lateral Muscles of the Lower Leg

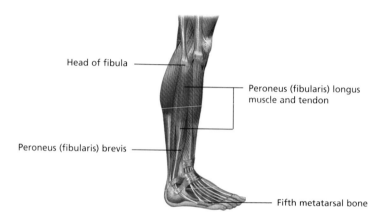

Head of fibula

Peroneus (fibularis) longus muscle and tendon

Peroneus (fibularis) brevis

Fifth metatarsal bone

The two muscles of the lateral (outer) compartment lie alongside the smaller of the two lower leg bones, the fibula. The peroneus (fibularis) longus is longer and lies more superficially. It arises from the upper part of the fibula and runs down to the sole of the foot. The peroneus (fibularis) brevis lies underneath the peroneus longus muscle. It arises from the lower part of the fibula and has a tendon that runs down to insert into the base of the fifth metatarsal bone of the foot. These muscles cause the foot to plantar flex (when the toes point down) and to evert (when the sole faces outward). They also help support the ankle by resisting the movement of inversion (sole facing inward), which is when the joint is weakest.

Body system:	musculoskeletal system
Location:	lateral (outer) compartment of the lower leg
Function:	enable plantar flexion and eversion of the foot, resist inversion of the ankle
Components:	peroneus (fibularis) longus, peroneus (fibularis) brevis
Related parts:	fibula, bones of the foot

The Superficial Calf Muscles

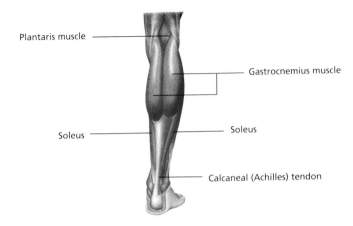

Plantaris muscle

Gastrocnemius muscle

Soleus

Soleus

Calcaneal (Achilles) tendon

The calf muscles are divided into superficial and deep layers. The largest and most superficial muscle is the gastrocnemius, which has a distinctive shape with two heads that arise from the condyles of the femur. The soleus is a large and powerful muscle that lies under the gastrocnemius and is important for maintaining balance when standing. The gastrocnemius and the soleus have a single common tendon known as the calcaneal (Achilles) tendon that runs from the lower calf to the heel. The tiny plantaris muscle is the weakest calf muscle and is sometimes absent. Together, these muscles have the job of plantar flexing the foot (lifting the toes and pointing the heels downward) and are vital during walking, running, and jumping.

Body system:	musculoskeletal system
Location:	posterior compartment of the lower leg
Function:	plantar flex the foot (heel lifted and toes pointing downward), vital in running and jumping and maintaining balance
Components:	gastrocnemius, soleus, plantaris
Related parts:	femur, tibia, fibula, calcaneus (heelbone)

The Deep Calf Muscles

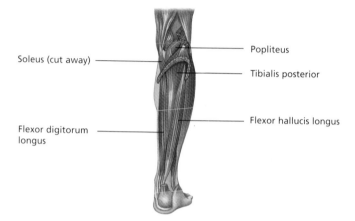

Soleus (cut away)

Popliteus

Tibialis posterior

Flexor hallucis longus

Flexor digitorum longus

Four muscles make up the deep group of calf muscles, the actions of which vary. The popliteus is a thin triangular muscle that lies at the back of the knee in the popliteal fossa. This muscle has the role of "unlocking" the knee joint by rotating it slightly to allow the straightened leg to be bent. The flexor digitorum longus has long tendons that pass down to the outer four toes to allow them to curl under. Although the flexor hallucis longus only runs to one toe—the big toe—it is a very powerful muscle that provides a "push off" or "spring" to the step during walking or running. The deepest muscle in the group is the tibialis posterior, the main provider of the action of inversion, in which the foot moves so that the sole faces inward.

Body system:	musculoskeletal system
Location:	posterior compartment of the lower leg
Function:	"unlock" the knee joint, curl toes, invert foot, provide the "push off" when walking or running
Components:	popliteus, flexor digitorum longus, flexor hallucis longus, tibialis posterior muscles
Related parts:	femur, tibia, fibula, bones of the foot

Arteries of the Leg

The lower limb is supplied with blood rich in oxygen and nutrients by a network of arteries that arise from the external iliac artery in the pelvis. The main artery of the leg is the femoral artery, which is a continuation of the iliac artery as it passes under the inguinal ligament in the groin. The main branch of the femoral artery is the profunda femoris, which itself supplies several branches, including the medial and lateral circumflex femoral arteries and the four perforating arteries. The popliteal artery is the continuation of the femoral artery from behind the knee. It runs down the back of the knee, with small arteries branching off to nourish that joint, before dividing into the anterior and posterior tibial arteries, which supply the structures and tissues in the anterior and posterior compartments of the lower leg and the foot.

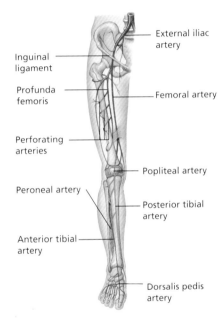

External iliac artery

Inguinal ligament

Profunda femoris

Perforating arteries

Peroneal artery

Anterior tibial artery

Femoral artery

Popliteal artery

Posterior tibial artery

Dorsalis pedis artery

Body system:	cardiovascular system
Location:	extending from the external iliac artery in the pelvis to the foot
Function:	provide blood rich in oxygen and nutrients to the tissues of the lower limb
Components:	external iliac, femoral, profunda femoris, perforating, popliteal, peroneal, posterior and anterior tibial, dorsalis pedis arteries
Related parts:	structures and tissues in the lower limb

Arteries Around the Knee

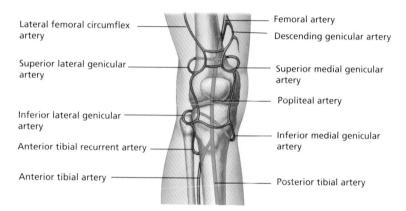

Lateral femoral circumflex artery

Femoral artery

Descending genicular artery

Superior lateral genicular artery

Superior medial genicular artery

Popliteal artery

Inferior lateral genicular artery

Anterior tibial recurrent artery

Inferior medial genicular artery

Anterior tibial artery

Posterior tibial artery

Behind the knee, the popliteal artery supplies a number of small branches that surround the knee joint and form connections, or anastomoses, with other small branches of the femoral and the anterior and posterior tibial arteries. Together they form an arterial network through which blood can bypass the normal route via the main popliteal artery. This is important when the knee is bent for a considerable time or if the main artery is narrowed or blocked. Just as the pulsations of the femoral artery can be felt in the groin, those of the popliteal artery can be felt behind the knee. However, since the popliteal artery lies so deeply within the tissues behind the knee, the pulse can sometimes be difficult to feel.

Body system:	cardiovascular system
Location:	extend from the femoral artery to the posterior and anterior tibial arteries
Function:	provide blood rich in oxygen and nutrients to the tissues of the lower limb
Components:	femoral, superior lateral, and medial genicular, popliteal, inferior lateral, and medial genicular, anterior and posterior tibial, etc.
Related parts:	structures and tissues in the lower limb

Arteries of the Foot

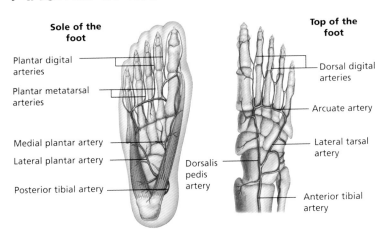

Sole of the foot

- Plantar digital arteries
- Plantar metatarsal arteries
- Medial plantar artery
- Lateral plantar artery
- Posterior tibial artery

Dorsalis pedis artery

Top of the foot

- Dorsal digital arteries
- Arcuate artery
- Lateral tarsal artery
- Anterior tibial artery

The small arteries in the foot form arches that interconnect, supplying branches to each side of the toes. Arterial blood is supplied by the terminal branches of the anterior and posterior tibial arteries. As the anterior tibial artery passes down in front of the ankle, it becomes the "dorsalis pedis" artery, which runs across the top of the foot toward the space between the first and second toes, where it joins the arteries on the sole of the foot. Branches of the dorsalis pedis on top of the foot join to form an arch that gives off branches to the toes. The sole of the foot has a rich blood supply provided by the branches of the posterior tibial artery. As this artery enters the sole, it divides to form the medial and lateral plantar arteries.

Body system:	cardiovascular system
Location:	branching through the foot
Function:	provide blood rich in oxygen and nutrients to the tissues of the foot
Components:	posterior and anterior tibial, lateral and medial plantar, plantar and dorsal digital, plantar and dorsal metatarsal, dorsalis pedis, etc.
Related parts:	structures and tissues in the foot

Saphenous Veins

There are two main superficial veins in the leg, the great and small saphenous veins. The great saphenous vein is the longest vein in the body, arising from the inner end of the dorsal venous arch of the foot and running up the leg toward the groin. On its journey, the great saphenous vein passes in front of the medial malleolus (inner ankle bone), tucks behind the medial condyle of the femur at the knee, and eventually drains into the femoral vein in the groin. The smaller saphenous vein arises from the lateral end of the dorsal venous arch and passes behind the lateral malleolus (outer ankle bone) and up the center of the back of the calf. At the knee, this smaller vein empties into the deep popliteal vein. The great and small saphenous veins receive blood along the way from many smaller veins and also intercommunicate with each other.

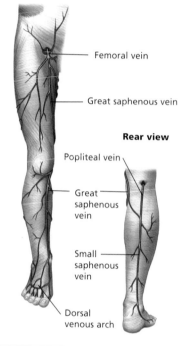

Femoral vein

Great saphenous vein

Rear view

Popliteal vein

Great saphenous vein

Small saphenous vein

Dorsal venous arch

Body system:	cardiovascular system
Location:	running from the dorsal arch of the foot to the femoral vein in the groin
Function:	drain deoxygenated blood from the tissues of the lower limb and return it to the heart
Components:	great and small saphenous veins
Related parts:	structures and tissues of the lower limb

Deep Veins of the Lower Limb

The deep veins of the lower limb follow the same pattern as the arteries, which they accompany along their length. As well as draining venous blood from the tissues of the limb, the deep veins also receive blood from the superficial veins via small connecting veins called perforating veins. Although the deep leg veins are referred to and illustrated as single veins, many of them are actually paired, one lying on either side of the artery. These paired veins are known as venae comitantes and are common throughout the body. The main deep veins of the lower limb are the posterior tibial (not shown), the anterior tibial, the popliteal (not shown), and the femoral veins. The large femoral vein, which is the continuation of the popliteal vein, receives blood from the superficial veins and continues up into the groin to become the external iliac vein.

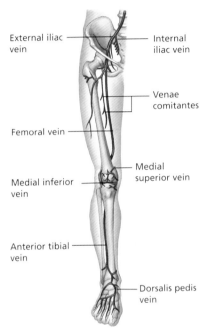

External iliac vein

Internal iliac vein

Venae comitantes

Femoral vein

Medial superior vein

Medial inferior vein

Anterior tibial vein

Dorsalis pedis vein

Body system:	cardiovascular system
Location:	running from the dorsalis pedis vein of the foot to the external iliac vein in the groin
Function:	drain deoxygenated blood from the tissues of the lower limb and return it to the heart
Components:	anterior and posterior tibial, popliteal and femoral veins
Related parts:	structures and tissues of the lower limb

Valves and the Venous Pump

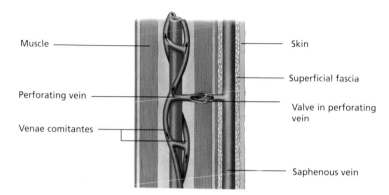

Muscle

Skin

Superficial fascia

Perforating vein

Valve in perforating vein

Venae comitantes

Saphenous vein

The arrangement of blood vessels in the leg means that blood flows from the superficial veins through the perforating veins to the deep veins. Venous blood is then moved back up through the body against gravity, helped toward the heart by the massaging action of the calf muscles that surround these deep veins (the venous pump). In addition, veins contain tiny valves that prevent backflow of the blood within them. These valves are of great importance in the veins of the lower limb. They ensure that, when the calf muscles contract, the blood is pushed up the vein toward the heart rather than back into the superficial veins. If the valves in the perforating veins become damaged, then backflow of blood can occur.

Body system:	cardiovascular system
Location:	veins in the lower limb
Function:	return deoxygenated blood to the heart from the tissues of the lower limb
Components:	vessels, valves
Related parts:	calf muscles

The Sciatic Nerve

Running from the thigh to the foot, the sciatic nerve supplies most of the leg muscles and is the largest nerve in the body. It is actually made up of two nerves—the tibial and the common peroneal (fibular) nerves—which are bound together by connective tissue to form a wide band that runs the full length of the back of the thigh. The sciatic nerve arises from a network of nerves at the base of the spine, called the sacral plexus, and passes out of the pelvis through an opening called the greater sciatic foramen. It curves under the gluteus maximus and down through the thigh, branching off into the hamstring muscles, to just above the knee where the two nerves usually separate to supply different parts of the leg. Sometimes the level at which the nerves divide varies, and rarely they emerge at the greater sciatic foramen as two separate nerves.

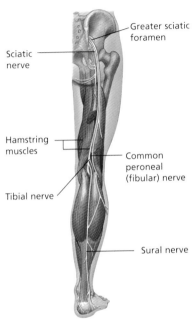

Greater sciatic foramen

Sciatic nerve

Hamstring muscles

Common peroneal (fibular) nerve

Tibial nerve

Sural nerve

Body system:	nervous system
Location:	extends from the sacral plexus in the spine to the foot
Function:	provides nerve supply to the muscles and other structures in the lower limb
Components:	common peroneal (fibular) nerve, tibial nerve
Related parts:	sacral plexus, muscles and tissues of the lower limb

The Common Peroneal Nerve

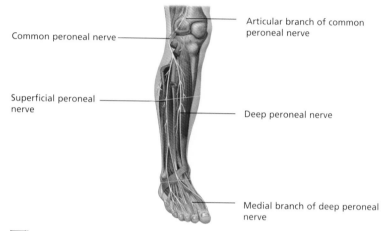

Articular branch of common peroneal nerve

Common peroneal nerve

Superficial peroneal nerve

Deep peroneal nerve

Medial branch of deep peroneal nerve

The common peroneal nerve leaves the sciatic nerve in the lower third of the thigh and runs down around the outer side of the lower leg before itself dividing into two just below the knee. The superficial branch of the peroneal nerve supplies the lateral (outer) compartment of the lower leg in which it lies. The deep peroneal nerve runs in front of the interosseous membrane, which connects the tibia and fibula, and then passes over the ankle into the foot. These two branches also supply the knee joint and the skin over the outer side of the calf and the top of the foot. As the common peroneal nerve passes around the outer side of the lower leg, it lies just under the skin and very close to the head of the fibula.

Body system:	nervous system
Location:	travels from the sacral plexus as the sciatic nerve, divides at the knee and extends down to the foot
Function:	provides nerve supply to the tissues of the lower leg
Components:	common peroneal nerve, superficial branch and articular branch of the peroneal nerve, deep peroneal nerve
Related parts:	sciatic nerve, structures and tissues of the lower limb

The Tibial Nerve

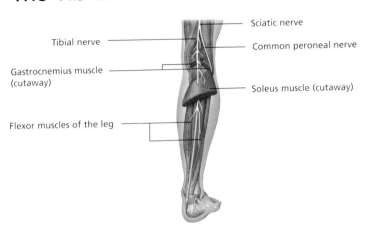

Sciatic nerve

Tibial nerve

Common peroneal nerve

Gastrocnemius muscle
(cutaway)

Soleus muscle (cutaway)

Flexor muscles of the leg

The larger of the two terminal branches of the sciatic nerve, the tibial nerve, supplies the flexors of the leg—those muscles that bend rather than straighten the joints. The tibial nerve arises in the lower third of the thigh, where it supplies the hamstring muscles. It then separates from the common peroneal nerve before following a course down the back of the leg under the large gastrocnemius and soleus muscles. At the ankle, the nerve passes behind the medial malleolus before dividing into the medial and lateral plantar nerves of the foot. The tibial nerve also has two cutaneous branches that supply areas of skin, the sural nerve (in the calf), and the medial calcaneal nerve (in the heel).

Body system:	nervous system
Location:	travels from the sacral plexus as the sciatic nerve, divides at the knee, and extends down to the foot
Function:	provides nerve supply to the tissues of the lower leg
Components:	tibial nerve, sural nerve, medial calcaneal nerve
Related parts:	sciatic nerve, structures and tissues of the lower limb

The Ankle

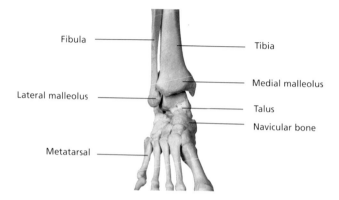

Fibula

Tibia

Medial malleolus

Lateral malleolus

Talus

Navicular bone

Metatarsal

At the ankle joint, a deep socket is formed by the lower ends of the tibia and fibula, the bones of the lower leg. Into this socket fits the pulley-shaped upper surface of the talus (ankle bone). The shape of the bones and the strength of the surrounding ligaments mean that the ankle is very stable, an important feature for such a major weightbearing joint. The articular surfaces (those parts that move against each other) of the ankle joint are covered with a layer of smooth hyaline cartilage. This cartilage is surrounded by synovial membrane that secretes a viscous fluid and helps to lubricate the joint. The ankle is a hinge joint and allows for dorsiflexion (toes up, heel down) and plantarflexion (toes down, heel up).

Body system:	musculoskeletal system
Location:	base of the tibia and fibula
Function:	bears the weight of the body, allows for movement of the foot in two planes—dorsiflexion and plantarflexion
Components:	undersurface of tibia, lateral malleolus of fibula, medial malleolus of tibia, trochlea of talus
Related parts:	bones of the foot

Ligaments of the Ankle

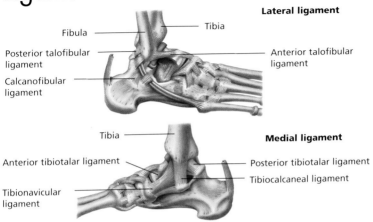

Lateral ligament

Fibula

Tibia

Posterior talofibular ligament

Anterior talofibular ligament

Calcanofibular ligament

Tibia

Medial ligament

Anterior tibiotalar ligament

Posterior tibiotalar ligament

Tibiocalcaneal ligament

Tibionavicular ligament

The ankle needs to be stable because it bears the entire weight of the body. The presence of a variety of strong ligaments around the ankle helps maintain this stability, while still allowing the necessary freedom of movement. Like most joints, the ankle is enclosed within a tough fibrous capsule, which is reinforced on each side by the medial (inner) and lateral (outer) ankle ligaments. Also known as the deltoid ligament, the medial ligament is a very strong structure that fans out from the tip of the medial malleolus of the tibia; each part of the ligament is named after the two bones that it connects. The lateral ligament is the weaker of the two, and it is made up of three distinct bands of fibrous tissue.

Body system:	musculoskeletal system
Location:	surrounding the ankle joint
Function:	strengthen the ankle joint, while allowing a wide range of movement
Components:	posterior talofibular, anterior talofibular, calcanofibular, anterior tibiotalar, tibionavicular, posterior tibiotalar, tibiocalcaneal
Related parts:	bones of the ankle and foot

The Tarsal Bones

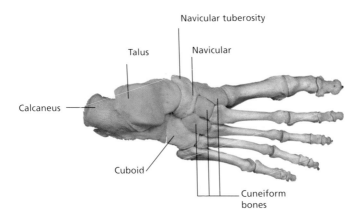

Navicular tuberosity

Talus

Navicular

Calcaneus

Cuboid

Cuneiform
bones

The human foot has 26 bones in total, seven of which are tarsal bones. The talus articulates with the tibia and fibula at the ankle joint and bears the full weight of the body. It lies above the calcaneus, or heel bone, and is the largest of the tarsal bones. The navicular is a relatively small bone with a projection (the navicular tuberosity), which is named for its boatlike appearance. Beneath the navicular is the cuboid, a bone that is roughly the shape of a cube. It lies on the outer side of the foot, and it has a groove on its undersurface to allow for the passage of a muscle tendon. The three cuneiforms are named according to their positions: medial, intermediate, and lateral. The medial is the largest of these wedge-shaped bones.

Body system:	musculoskeletal system
Location:	between the tibia and the fibula and the metatarsals
Function:	provide structure to the foot, enabling movement; bear the weight of the body and aid balance
Components:	talus, calcaneus, navicular, cuneiforms, cuboid
Related parts:	tibia and fibula, bones of the foot

The Calcaneus

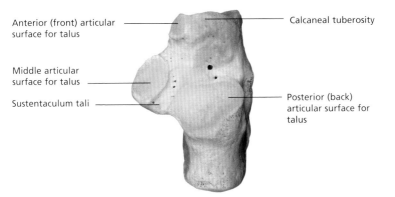

Anterior (front) articular surface for talus

Calcaneal tuberosity

Middle articular surface for talus

Sustentaculum tali

Posterior (back) articular surface for talus

The calcaneus is the largest bone in the foot and can easily be felt under the skin as the heel. It is a large, strong bone, with the important role of transmitting the weight of the body from the talus to the ground. The talus has articular surfaces where it forms joints with the talus above and the cuboid in front. The inner surface of the calcaneus bears a projection, the sustentaculum tali, which supports the head of the talus. On the underside of this projection is a groove for the passage of a long tendon. The back of the calcaneus has a roughened prominence, the calcaneal tuberosity, which comes into contact with the ground when standing. Halfway up the talus is a ridge, which indicates the site of attachment of the Achilles tendon.

Body system:	musculoskeletal system
Location:	below the talus at the rear of the foot
Function:	provides structure to the foot, enabling movement; bears the weight of the body and aids balance
Components:	articular surfaces, calcaneal tuberosity, sustentaculum tali
Related parts:	bones of the foot

Metatarsals and Phalanges

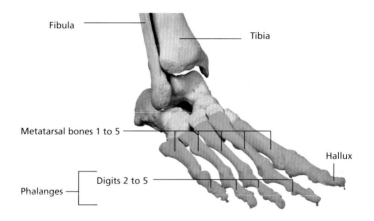

Fibula

Tibia

Metatarsal bones 1 to 5

Hallux

Digits 2 to 5

Phalanges

The metatarsals and phalanges in the foot are miniature long bones (consisting of a base, a shaft, and a head), which provide stability for the foot. The bases of the metatarsals articulate with the tarsal bones in the middle of the foot and the heads articulate with the phalanges of the corresponding toes. The metatarsals are numbered one to five, starting with the most medial, which lies behind the big toe. The first metatarsal is shorter and sturdier than the rest and articulates with the first phalanx of the big toe. There are 14 phalanges in the foot, which are smaller and less mobile than those of the fingers, but they have the same arrangement. Each toe has three phalanges, except for the big toe (hallux) which has only two.

Body system:	musculoskeletal system
Location:	extend from the tarsal bones
Function:	provide structure to the foot, enabling movement; bear the weight of the body, and aid balance.
Components:	metatarsal bones 1–5, phalanges 1– 14
Related parts:	tarsal bones, tibia, and fibula

Ligaments of the Foot

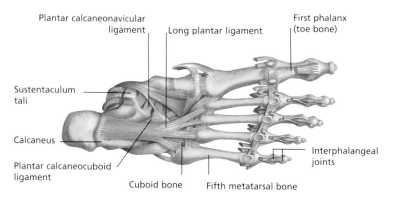

Plantar calcaneonavicular ligament

Long plantar ligament

First phalanx (toe bone)

Sustentaculum tali

Calcaneus

Plantar calcaneocuboid ligament

Interphalangeal joints

Cuboid bone

Fifth metatarsal bone

View of the underside of the foot

The bones of the foot are arranged in such a way that they form bridgelike arches, which allows the foot to be flexible enough to cope with uneven ground while bearing the weight of the body. The foot arches are supported by the presence of a number of ligaments that lie on the plantar (under) surface of the bones and provide a firm but flexible base. There are three major ligaments: the plantar calcaneonavicular ligament, the long plantar ligament, and the plantar calcaneocuboid ligament. Many other ligaments support and bind together the long metatarsals and the phalanges (toe bones). The metatarsals are bound to the tarsals and to each other by ligaments running across the foot on both their surfaces.

Body system:	musculoskeletal system
Location:	on the plantar (under) surface of the foot
Function:	provide a firm flexible base that supports the arches of the foot
Components:	plantar calcaneonavicular, long plantar, plantar calcaneocuboid ligaments
Related parts:	bones of the foot

Joints of the Foot

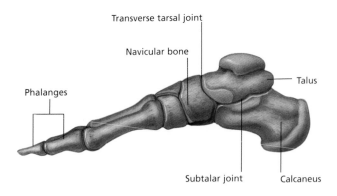

Transverse tarsal joint

Navicular bone

Talus

Phalanges

Subtalar joint

Calcaneus

The ankle joint allows the foot to move only up and down. Other movements of the foot, such as eversion, in which it faces outward, or inversion, where it turns inward, take place further down the foot at two joints: the transverse tarsal and the subtalar joints. The transverse tarsal is a complicated joint formed by the adjoining articular surfaces of parts of the calcaneus (heel bone), talus (ankle bone), navicular and cuboid bones. The subtalar joint is formed where the talus moves against the calcaneus. There are many other small synovial (fluid-filled) joints located in the foot where bone meets bone. However, these are generally held tightly together by tough ligaments and so little movement is possible.

Body system:	musculoskeletal system
Location:	between the bones of the foot
Function:	allow movement of the foot while maintaining stability
Components:	transverse tarsal and subtalar joints
Related parts:	bones, muscles, and ligaments of foot

Retinacula of the Ankle

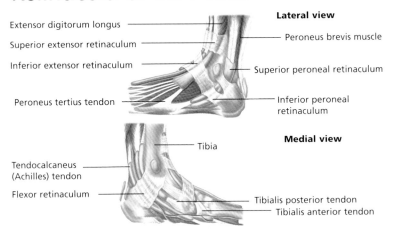

Lateral view

Extensor digitorum longus

Superior extensor retinaculum

Inferior extensor retinaculum

Peroneus tertius tendon

Peroneus brevis muscle

Superior peroneal retinaculum

Inferior peroneal retinaculum

Medial view

Tibia

Tendocalcaneus (Achilles) tendon

Flexor retinaculum

Tibialis posterior tendon

Tibialis anterior tendon

Many of the muscles that move the foot are those of the lower calf, and long tendons are necessary to extend to the bones of the foot. Where these tendons cross the ankle joint, they are held firmly in place by a series of fibrous bands, or retinacula. There are four main retinacula: the superior and inferior extensor retinacula, both of which retain the extensor muscles; the peroneal retinaculum, which holds the peroneal muscle tendons in place; and the flexor retinaculum, which retains the long flexor tendons. As the muscles of the lower leg move the foot and toes, the long tendons run back and forth against the bones of the ankle. To prevent friction, they are enclosed in a synovial "sheath" lubricated with fluid.

Body system:	musculoskeletal system
Location:	around the ankle and upper foot
Function:	hold the muscle tendons in place, providing stability to the foot
Components:	superior and inferior extensor, peroneal and flexor retinacula
Related parts:	muscles and bones of the lower leg and foot

Muscles on the Top of the Foot

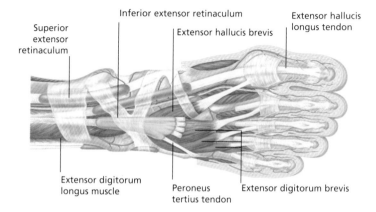

Superior extensor retinaculum

Inferior extensor retinaculum

Extensor hallucis brevis

Extensor hallucis longus tendon

Extensor digitorum longus muscle

Peroneus tertius tendon

Extensor digitorum brevis

Most of the muscles that lie within the foot, the intrinsic muscles, are in the sole; the top, or dorsal, surface of the foot has just two muscles. The extensor digitorum brevis and extensor hallucis brevis are, as their names suggest, short muscles that extend (straighten or pull upward) the first four toes. The extensor digitorum brevis arises from the upper surface of the calcaneus (heel bone) and the inferior extensor retinaculum, and it divides into three parts, each of which has a tendon that inserts into the second, third, or fourth toes. The extensor hallucis brevis is really part of the extensor digitorum brevis and runs down to insert into the big toe, or "hallux" from which its name derives.

Body system :	musculoskeletal system
Location:	on the dorsal surface of the foot, extending to the first four toes
Function:	extend (straighten) the first four toes
Components:	extensor digitorum brevis, extensor hallucis brevis
Related parts:	bones, muscles, and tendons of the foot

Muscles of the Sole

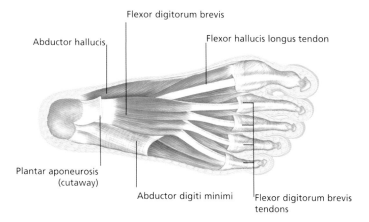

Flexor digitorum brevis

Abductor hallucis

Flexor hallucis longus tendon

Plantar aponeurosis
(cutaway)

Abductor digiti minimi

Flexor digitorum brevis
tendons

The sole of the foot has four layers of intrinsic muscles, which work with the extrinsic muscles (those in the lower leg) to meet the varying demands placed upon the foot. The most superficial layer of sole muscles is located just under a thick sheet of connective tissue called the plantar aponeurosis. The abductor hallucis muscle lies on the inner border of the sole and acts to abduct (move outward) and flex (straighten) the big toe. The flexor digitorum brevis is a fleshy muscle in the center of the sole whose four tendons insert into each of the lateral four toes. Contraction of this muscle causes the toes to flex. Lying along the outer side of the sole is the abductor digiti minimi, which acts to abduct and flex the little toe.

Body system:	musculoskeletal system
Location:	under the plantar aponeurosis in the sole of the foot
Function:	abduct and flex the toes
Components:	abductor hallucis, flexor digitorum brevis, abductor digiti minimi
Related parts:	bones, muscles, and tendons of the foot

The Skeleton

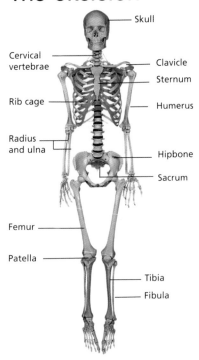

Skull

Cervical vertebrae

Clavicle

Sternum

Rib cage

Humerus

Radius and ulna

Hipbone

Sacrum

Femur

Patella

Tibia

Fibula

The skeleton is made up of bone and cartilage and accounts for about one-fifth of the body's weight. The skull, the vertebral column, and the rib cage are known as the axial skeleton, and the bones of the limbs, together with the pectoral and pelvic girdles, form the appendicular skeleton. The skeleton has a number of vital functions. It provides a supportive framework for the body and holds the soft internal organs in place. The brain and spinal cord are protected by the skull and vertebral column, while the rib cage protects the heart and lungs. Throughout the body, muscles attach to the bones to give them leverage to bring about movement. Minerals, such as calcium and phosphate, are stored in bones, and blood cells are manufactured in the marrow cavity of some bones.

Body system:	musculoskeletal system
Location:	extends from the skull to the toes
Function:	provides framework for the rest of the body, protects vital organs, enables movement by providing sites of attachment for muscles, stores minerals, manufactures blood cells
Components:	more than 200 separate bones, cartilage
Related parts:	organs and tissues of the body

Types of Joints

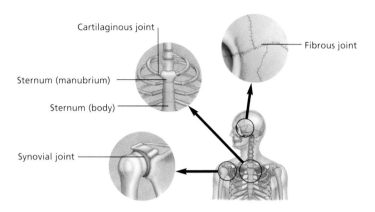

Cartilaginous joint

Fibrous joint

Sternum (manubrium)

Sternum (body)

Synovial joint

A joint is formed where two or more bones meet, allowing movement and providing support. The joints of the body can be divided into three main structural groups—fibrous, cartilaginous, and synovial. Bones connected by a fibrous joint, for example those of the skull, are held together by a protein called collagen, which allows little, if any, movement. Cartilaginous joints also prevent movement but can "relax" under pressure allowing flexibility. The ends of the bones in cartilaginous joints, such as those between the sternum and ribs, are covered with hyaline cartilage and connected by tough fibrocartilage. Synovial joints allow easy movement between bones and contain fluid that lubricates the joint.

Body system :	musculoskeletal system
Location:	between bones
Function:	connect bones, sometimes allowing movement and flexibility
Components:	fibrous, cartilaginous, synovial joints
Related parts:	bones, ligaments, muscles

Skeletal Muscle

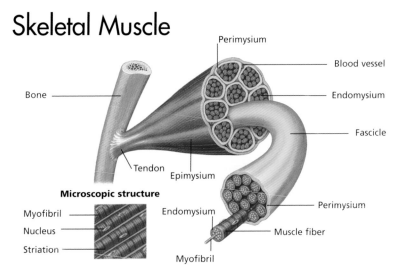

Microscopic structure

Perimysium

Blood vessel

Bone

Endomysium

Fascicle

Tendon

Epimysium

Myofibril

Nucleus

Endomysium

Perimysium

Striation

Muscle fiber

Myofibril

The most familiar muscles in the body are skeletal muscles (also known as voluntary or striated), many of which are visible under the skin. Skeletal muscles are responsible for movement and are under voluntary control, although they can also contract involuntarily in a reflex action. The fibers of skeletal muscle are bound together by connective tissue (epimysium) and divided into groups or fascicles by a sheath (perimysium). Within these fascicles, each muscle fiber is surrounded by an endomysium. The whole muscle is attached to bone by a tough fibrous band, the muscle tendon. Skeletal muscle can contract powerfully, exerting a great deal of force, or it can enable fine, delicate movements.

Body system:	musculoskeletal system
Location:	throughout the body
Function:	enables voluntary movement
Components:	tendon, epimysium, perimysium, endomysium, fascicles, fibers
Related parts:	bones, ligaments

Smooth (Involuntary) Muscle

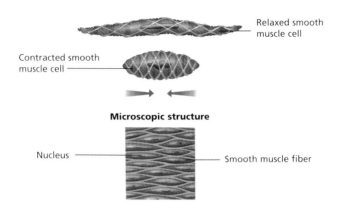

Relaxed smooth muscle cell

Contracted smooth muscle cell

Microscopic structure

Nucleus

Smooth muscle fiber

Smooth muscle is so named because of its lack of striations, or stripes, when viewed under the microscope. It is also known as involuntary muscle because its actions do not come under an individual's conscious control but are regulated by the autonomic nervous system. Smooth muscle is found in the walls of hollow structures within the body, such as the gut, blood vessels, and bladder. Here, it acts to regulate the size of the lumen (central space), as well as causes wavelike contractions (peristalsis) in some organs (such as the intestines and ureters). Smooth muscle is also found in the skin, where it acts upon hairs, and in the eyeball, where it determines the thickness of the lens and the size of the pupil.

Body system:	musculoskeletal system
Location:	gut, blood vessels, bladder, ureters, skin, eye, etc.
Function:	regulates the size of the central space of organs, enables peristalsis, alters thickness of lens and size of pupil
Components:	smooth muscle fibers
Related parts:	autonomic nervous system

Arterial Circulation

The role of the systemic arterial system is to transport blood from the heart to nourish the tissues of the body. Oxygenated blood from the lungs is first pumped into the aorta (the main artery of the body) via the heart. Branches from the aorta pass to the head, upper limbs, trunk, and lower limbs in turn. These large branches supply smaller branches, which divide repeatedly, until they become tiny arterioles that supply the tissues and organs of the body. The walls of the arteries are elastic and muscular to enable them to withstand the high pressure as the heart pumps blood into the arterial system. Most arteries lie fairly deep within the body to protect them from damage; if an artery is severed, a great deal of blood can be lost in a short space of time because of the high pressure within the vessel.

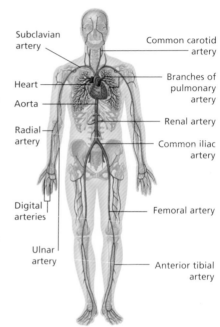

Subclavian artery

Common carotid artery

Heart

Branches of pulmonary artery

Aorta

Radial artery

Renal artery

Common iliac artery

Digital arteries

Femoral artery

Ulnar artery

Anterior tibial artery

Body system:	cardiovascular system
Location:	throughout the body
Function:	supplies all the organs and tissues of the body with nutrients and oxygen
Components:	numerous arteries and arterioles
Related parts:	heart, venous circulation

The Venous System

The systemic venous system carries blood back to the heart from the tissues. This blood is then pumped through the pulmonary circulation to be reoxygenated before entering the systemic circulation again. Veins originate in tiny venules that receive blood from the capillaries. The veins converge on one another, forming increasingly larger vessels until the two main collecting veins of the body, the superior and inferior vena cavae, are formed. These two large veins drain directly into the heart. At any one time about 65 percent of the total blood volume is contained in the venous system. Veins have thinner walls than arteries and are more superficial in the body. In addition, there is no pumping mechanism to force blood around the system, and the veins, therefore, need built-in valves to prevent the backflow of blood.

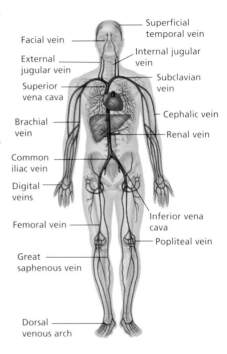

Facial vein
Superficial temporal vein
Internal jugular vein
External jugular vein
Subclavian vein
Superior vena cava
Cephalic vein
Brachial vein
Renal vein
Common iliac vein
Digital veins
Inferior vena cava
Femoral vein
Popliteal vein
Great saphenous vein
Dorsal venous arch

Body system :	cardiovascular system
Location:	throughout the body
Function:	returns deoxygenated blood to the heart from the organs and tissues of the body
Components:	numerous veins and venules
Related parts:	arterial system, heart

Pulmonary Circulation

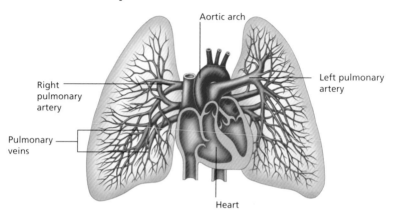

The pulmonary circulation allows blood to come into close contact with the alveoli (air sacs) of the lungs so that oxygen can pass into the blood and carbon dioxide can pass out into the lungs for excretion. With each beat of the heart, deoxygenated blood is pumped from the right ventricle into the lungs through the right and left pulmonary arteries. After many arterial divisions, the blood flows through the dense network of capillaries that surrounds the alveoli (air sacs) of the lung and is reoxygenated. This freshly oxygenated blood eventually enters the four pulmonary veins, which return to the left atrium of the heart. From here the blood is pumped out to the arterial systemic circulation to nourish the tissues.

Body system:	cardiovascular system
Location:	vessels extend from the heart to the lungs
Function:	enable blood to come into contact with the functioning structures of the lung—the alveoli—to allow gaseous exchange to take place
Components:	pulmonary arteries, pulmonary veins
Related parts:	heart, alveoli

The Peripheral Nervous System

The nervous system of the body is divided into two parts: the central nervous system (CNS), consisting of the brain and spinal cord, and the peripheral nervous system (PNS). The PNS has three major components: sensory receptors, which are specialized nerve endings that receive information, for example, about temperature, pain, and touch; motor nerve endings, which cause the muscle on which they lie to contract in response to a signal from the CNS; and peripheral nerves—bundles of nerve fibers that carry information to and from the CNS. Some of the peripheral nerves (the 12 pairs of cranial nerves) emerge directly from the brain and are concerned with receiving information from, and enabling control of, the head and neck. The spinal nerves arise from the spinal cord and contain thousands of nerve fibers that supply the rest of the body.

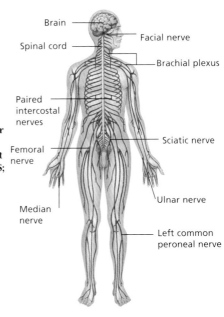

Brain

Spinal cord

Facial nerve

Brachial plexus

Paired intercostal nerves

Femoral nerve

Sciatic nerve

Median nerve

Ulnar nerve

Left common peroneal nerve

Body system:	nervous system
Location:	cranial nerves emerge from the brain stem, spinal nerves arise from the spinal cord
Function:	enable control of the head and neck, provide sensory and motor nerve supply to the body
Components:	12 pairs of cranial nerves, 31 pairs of spinal nerves
Related parts:	brain, spinal cord

Neurons

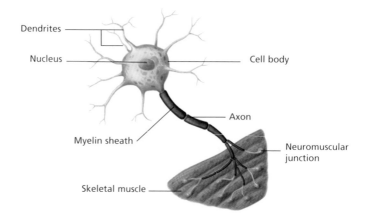

Dendrites

Nucleus

Cell body

Axon

Myelin sheath

Neuromuscular junction

Skeletal muscle

The tissues of the nervous system are made up of two types of cells: neurons, or nerve cells, which transmit information in the form of electrical signals; and glial cells, the smaller supporting cells that surround them. Neurons are the large, highly specialized cells of the nervous system whose function is to receive information and transmit it through the body. They all have a single cell body from which a number of dendrites, or branching processes, emerge. Each neuron also has a long axon that carries electrical impulses away from the cell body and is covered by an insulating myelin sheath. Neurons are unable to divide to replace themselves if they become damaged and, therefore, have a very long life.

Body system:	nervous system
Location:	throughout the body
Function:	transmit information in the form of electrical impulses
Components:	cell body, dendrites, axon, myelin sheath
Related parts:	tissues of the body, nervous system

Myelin Sheath

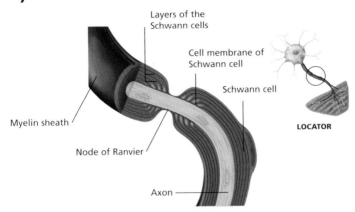

Layers of the
Schwann cells

Cell membrane of
Schwann cell

Schwann cell

Myelin sheath

LOCATOR

Node of Ranvier

Axon

The speed of an electrical signal along a neuron's axon is increased by the presence of a myelin sheath, a layer of fatty insulation. In the peripheral nervous system, the myelin sheath is produced by specialized Schwann cells that wrap themselves around the axon of a nerve cell to form a sheath of concentric circles. Each Schwann cell lies adjacent to, but not touching, the next. The gap between the cells, where there is no myelin, is known as the node of Ranvier. As an electrical impulse passes down the nerve, it must "hop" from one node to the other, which makes it travel faster overall than if no myelin sheath were present. In fact, myelinated nerve fibers can transmit information up to 150 times faster.

Body system:	nervous system
Location:	surrounding the axon of neurones
Function:	increase speed of transmission of electrical impulses along nerve fibers
Components:	fatty, insulating tissue
Related parts:	tissues of the nervous system

Peripheral Nerves

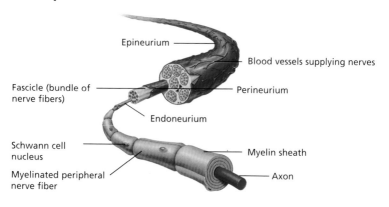

The greater part of each peripheral nerve is made up of three protective connective tissue coverings, without which the fragile nerve fibers would be vulnerable to injury. The endoneurium is a layer of delicate tissue that surrounds the smallest unit of the peripheral nerve, the axon. This layer may also enclose the axon's myelin sheath. The perineurium encloses a group, or fascicle, of protected nerve fibers. These fascicles are tied together in bundles and surrounded by a tough connective tissue coat, the epineurium, to form a peripheral nerve. The epineurium also encloses blood vessels that help to nourish and oxygenate the nerve fibers and their connective tissue coverings.

Body system:	nervous system
Location:	throughout the body
Function:	peripheral nerves carry information to and from the central nervous system
Components:	axon, myelin sheath, fascicles, endoneurium, epineurium, perineurium
Related parts:	tissues of the nervous system

The Autonomic Nervous System

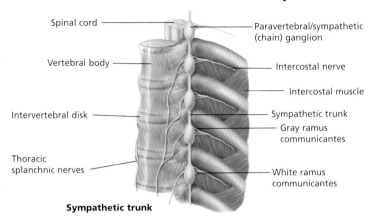

Spinal cord

Paravertebral/sympathetic (chain) ganglion

Vertebral body

Intercostal nerve

Intercostal muscle

Intervertebral disk

Sympathetic trunk

Gray ramus communicantes

Thoracic splanchnic nerves

White ramus communicantes

Sympathetic trunk

The autonomic nervous system provides the nerve supply to those parts of the body that are not consciously directed. It can be divided into two parts, the sympathetic and parasympathetic systems. In each system, two nerve cells make up the pathway from the central nervous system to the relevant organ. The cell bodies of the sympathetic nervous system lie within a section of the spinal cord, and fibers leave the cord via a chain of ganglia close to the cord. The sympathetic system has effects on the body that are often referred to as the "fight-or-flight" response. In dangerous situations, the system becomes more active, causing the heart rate to increase and the skin to become pale as blood is diverted to the muscles.

Body system:	nervous system
Location:	originates in central nervous system and extends to organs
Function:	sympathetic system provides nerve supply to those parts of the body that are not under conscious control; for example, it speeds up the heart rate and slows down the digestive process
Components:	nerve cell bodies and fibers
Related parts:	rest of the nervous system, organs

The Parasympathetic System

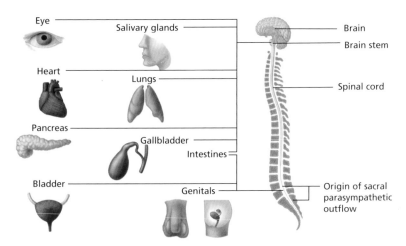

Eye
Salivary glands
Brain
Brain stem
Heart
Lungs
Spinal cord
Pancreas
Gallbladder
Intestines
Bladder
Genitals
Origin of sacral parasympathetic outflow

The parasympathetic system is the part of the autonomic nervous system most most active during periods of rest. Its structure is much simpler as the cell bodies of the first of the two neurons in the pathway are located in only two places. Fibers from the parasympathetic nervous system in the gray matter of the brain stem leave the skull as a number of cranial nerves. Together these fibers make up what is known as the cranial parasympathetic outflow. The sacral region of the spinal cord also contains parasympathetic fibers, which leave the cord through the ventral root. The parasympathetic nervous system supplies the same organs as the sympathetic, but has opposing effects.

Body system:	nervous system
Location:	originates in central nervous system and extends to organs
Function:	parasympathetic system provides nerve supply to those parts of the body that are not under conscious control; for example, it slows down the heart rate and speeds up the digestive process
Components:	nerve cell bodies and fibers
Related parts:	rest of the nervous system, organs

The Lymphatic System

The lymphatic system is the lesser known part of the circulatory system that works together with the cardiovascular system to transport a fluid called lymph around the body. Lymph is a clear watery fluid, which is derived from blood and bathes the body's tissues. It also contains lymphocytes, specialized white blood cells that attack and destroy foreign micro-organisms and, therefore, plays a vital role in the defense of the body against disease. The lymphatic system consists of a network of lymph vessels that runs throughout the body collecting fluid that has leaked from blood vessels. Collections of lymph nodes are scattered throughout the system along the routes of the vessels and these filter the lymph as it passes through. At various points in the body, for example, at the thoracic duct, lymph is drained back into the venous circulation.

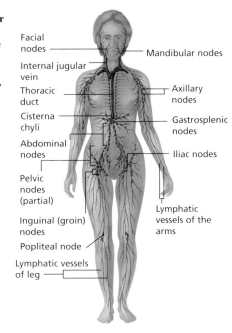

Facial nodes

Mandibular nodes

Internal jugular vein

Thoracic duct

Axillary nodes

Cisterna chyli

Gastrosplenic nodes

Abdominal nodes

Iliac nodes

Pelvic nodes (partial)

Inguinal (groin) nodes

Lymphatic vessels of the arms

Popliteal node

Lymphatic vessels of leg

Body system :	lymphatic system
Location:	throughout the body
Function:	drains excess fluid from the tissues of the body, filters out harmful bacteria, produces lymphocytes to fight infection
Components:	lymphatic vessels, nodes, lymphocytes, lymphoid tissues and organs
Related parts:	venous circulation

Lymph Node and Vessel

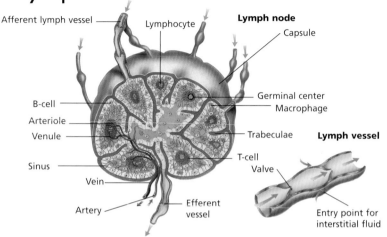

Afferent lymph vessel — Lymphocyte — **Lymph node** — Capsule — Germinal center — Macrophage — B-cell — Arteriole — Venule — Sinus — Vein — Artery — Efferent vessel — Trabeculae — T-cell — Valve — **Lymph vessel** — Entry point for interstitial fluid

L ymph nodes are small, round structures that lie along the course of the lymphatic vessels and act as filters of the lymph (fluid from around the body's cells). As well as fluid, the tiny lymphatic vessels in the tissues may pick up debris, such as cell particles, bacteria, and viruses. As the lymph passes through the node, it comes into contact with cells that ingest any solid particles and recognize foreign micro-organisms, preventing them from entering the bloodstream. Lymph vessels vary in size; tiny lymph capillaries receive fluid that has leaked out of the blood vessels. These capillaries join to form larger vessels that eventually converge to form the two lymphatic ducts—the thoracic duct and the right lymphatic duct.

Body system:	lymphatic system
Location:	throughout the body
Function:	drain excess fluid from the tissues of the body, filter out harmful bacteria, produce lymphocytes t o fight infection
Components:	fibrous capsule, trabeculae, sinus, germinal center, lymph and blood vessels
Related parts:	venous circulation

Lymphoid Tissues and Organs

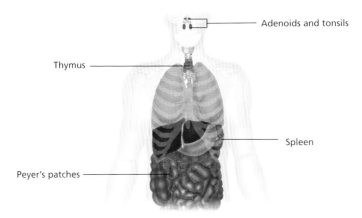

Adenoids and tonsils

Thymus

Spleen

Peyer's patches

Scattered throughout the body are discrete groups of lymphoid tissue, which play an important role in the immune system. The spleen provides a site for the cells of the immune system to proliferate and monitors the blood for foreign or damaged cells. The thymus is a small gland that lies in the chest just behind the upper part of the sternum (breastbone). It receives newly formed lymphocytes from the bone marrow, which mature into T-cells. Lymphoid tissue in the gastrointestinal tract lies just beneath the lining of the gut, at the back of the mouth in the form of tonsils, and as clumps of nodules known as Peyer's patches in the walls of the small intestine. These help to protect against infectious organisms that enter through the mouth.

Body system:	lymphatic system
Location:	throughout the body
Function:	provide sites for cells of the immune system to proliferate and mature, protect against infection
Components:	spleen, thymus, adenoids, tonsils, gut lining, Peyer's patches
Related parts:	blood circulation

The Endocrine System

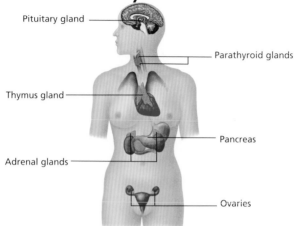

Pituitary gland

Parathyroid glands

Thymus gland

Pancreas

Adrenal glands

Ovaries

The endocrine system is formed of a collection of specialized glands scattered throughout the body. These glands secrete essential substances known as hormones (specialized proteins) directly into the surrounding tissue from where they are absorbed into the bloodstream. The major endocrine organs in the body include the pituitary gland in the brain, the thyroid and parathyroid glands, the thymus gland, the pancreas, the adrenal glands, which are located on top of the kidneys, and, in women, the ovaries. Each of these produces one or more hormones that are vital in regulating the body's function. For example, the pancreas secretes the hormone insulin, which regulates the body's use of sugar.

Body system:	endocrine system
Location:	throughout the body
Function:	secrete essential hormones that regulate the body's metabolism
Components:	pituitary, thymus, thyroid, parathyroids, pancreas, adrenals, ovaries
Related parts:	many other organs in the body, which the hormones act upon

The Skin

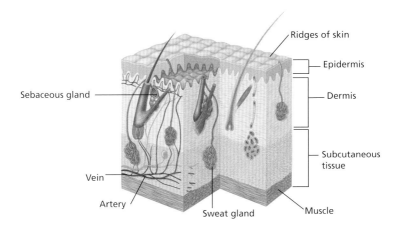

Ridges of skin

Epidermis

Dermis

Sebaceous gland

Subcutaneous
tissue

Vein

Artery

Sweat gland

Muscle

The skin covers the entire surface area of the body and has a surface area of about 1.5–2 yd² (1.5–2 m²). Skin is composed of two layers, the epidermis and the dermis. The epidermis is the thinner of the two layers and serves as a tough, protective covering. It is made up of layers of cells, the innermost of which divide rapidly, providing cells for the outer layers. The dermis is the thicker layer of skin, which lies protected under the epidermis. It is composed of connective tissue and has elastic and collagen fibers for suppleness and strength. The dermis contains a rich supply of blood vessels, as well as numerous sensory nerve endings. Lying within this layer are the hair follicles, and the sweat and sebaceous (oil) glands.

Body system :	integumentary system
Location:	covering the entire body
Function:	protection, heat regulation, prevention of dehydration and infection, provision of sensation
Components:	epidermis, dermis
Related parts:	hair, fat, muscle

Nails

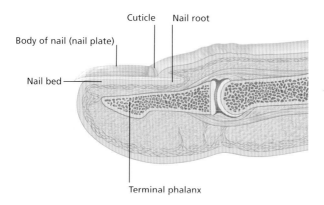

Cuticle · Nail root

Body of nail (nail plate)

Nail bed

Terminal phalanx

Nails form a hard protective covering for the vulnerable tips of the fingers and toes and provide a useful tool for scratching or scraping. They lie on the dorsal (back) surface of the digits, overlying the terminal phalanx (or final bone) of each. Each nail is composed of a plate of hard keratin that is continuously produced at its root, and except for the free edge at the top, the nail is surrounded and overlapped by skin folds. At the base of the nail, beneath the nail itself and the nail fold, is the root or matrix and it is here that the keratin is produced by cell division. The lunula is the pale crescent-shaped area located at the base of the nail, which is bordered by the cuticle, a protective fold of skin that protects the matrix from infection.

Body system:	integumentary system
Location:	on the dorsal side of each terminal phalanx
Function:	protect delicate ends of fingers and toes, act as a tool
Components:	matrix, nail bed, cuticle, nail plate
Related parts:	fingers and toes

Hair

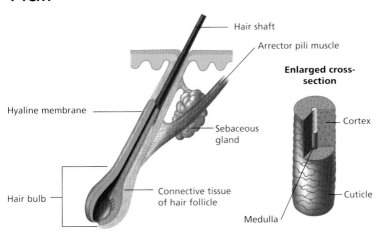

Hair shaft

Arrector pili muscle

Enlarged cross-section

Hyaline membrane

Cortex

Sebaceous gland

Hair bulb

Connective tissue of hair follicle

Cuticle

Medulla

The surface of the human body is covered with millions of hairs that are most noticeable on the head, around the external genitalia, and under the arms. Although human hair does not really act as insulation for the body, it can contribute to sensation, and it protects the eyes and scalp. Hair is composed of flexible strands of a protein, keratin, which is produced by follicles within the dermis of the skin. Each follicle has an expanded end—the bulb—which receives a knot of capillaries to nourish the root of the growing hair shaft. Sebaceous glands lie alongside the follicle and produce oil to lubricate the hair. A tiny muscle, the arrector pili, is able to contract, making the hair "stand on end" if stimulated by cold or fear.

Body system:	integumentary system
Location:	throughout the body
Function:	contributes to sensation, protects eyes, insulates scalp
Components:	follicle, hair shaft
Related parts:	skin, sweat glands, blood vessels

Index

319